EXPERIMENTS
in BIOREGIONALISM

Futures of New England

Experiments in Bioregionalism
The New England River Basins Story
Charles H. W. Foster, 1984

Working with Your Woodland
A Landowner's Guide
Mollie Beattie, Charles Thompson,
and Lynn Levine, 1983

New England Prospects
Critical Choices in a Time of Change
Carl H. Reidel, *Editor*, 1982

Wildlands and Woodlots
The Story of New England's Forests
Lloyd C. Irland, 1982

The New England Regional Plan
An Economic Development Strategy
New England Regional Commission, 1981

Business and Academia
Partners in New England's Economic Renewal
John C. Hoy and Melvin Bernstein, *Editors*, 1981

EXPERIMENTS
in BIOREGIONALISM

*The New England River
Basins Story*

Charles H. W. Foster

University Press of New England
HANOVER AND LONDON, 1984

University Press of New England

Brandeis University

Brown University

Clark University

Dartmouth College

University of New Hampshire

University of Rhode Island

Tufts University

University of Vermont

Printed in the United States of America

LIBRARY OF CONGRESS
CATALOGING IN PUBLICATION DATA

Foster, Charles H. W., 1927–
   Experiments in bioregionalism.

   Bibliography: p.
   Includes index.
   1. Watershed management—New England. 2. Water resources development—New England. I. Title.
II. Title: Bioregionalism.

TC423.15.F67   1984   333.91'02'0974   83-40554
ISBN 0-87451-301-4

Contents

Preface

There is abroad in the land a concept termed *bioregionalism*. It is defined in this context as natural resources–related issues and events that occur in transboundary settings. For years, we have known that resources and environmental matters of size and substance are rarely respecters of conventional boundaries and must be addressed at levels beyond those of fixed political jurisdictions. River basins, forest types, migratory fish and waterfowl, and air pollution are just a few examples of this principle. But too little attention has been paid to the bioregional institutions created for such purposes—their efficacy, viability, and longevity. In point of fact, an alarming number of such efforts seem to founder, usually on the rocky shoals of jurisdictional disagreement. Those that do survive remain invariably suspect, for bioregional managers must walk a constant tightrope between a constituency that insists on political currency, and the longer-range objectives their institution is supposed to address. If the operation is not professionally sound, it is in immediate trouble. If it is successful, it soon loses the attention of top leadership, which must necessarily direct its energies toward matters of crisis and moment. This leads, in time, to a delegation of functions lower and lower in the hierarchy and sows the seeds of growing dissatisfaction and eventual dissolution.

Viewed over time, bioregionalism has enjoyed a slim and inconclusive history in the United States. Its intellectual genesis is traceable to the early activists and spokesmen for the conservation movement. Witness John Muir and his preoccupation with the High Sierras, or Henry David Thoreau and his notable interest in such bioregions as Walden Pond, Cape Cod, or the Maine woods. Many of the larger concepts that now drive the environmental movement, and have had their origins in particular issues in particular places, are bioregional in character. The Great Plains dustbowl, for example, helped seed a national soil and water conservation program. The ravage of the New England forest brought about passage of the Weeks Act and the enabling legislation for the purchase of the eastern national forests.

But the real flowering of bioregionalism seems to have occurred in the 1920s and 1930s, sparked by the eloquence of social pioneers like

Lewis Mumford and Benton MacKaye, and the activism of the New Deal. The Tennessee Valley Authority (TVA) became one of the earliest tangible bioregional instruments. Agencies like the National Planning Board spawned some of the earliest natural resources planning institutions in regions like New England and the Pacific Northwest.

Following World War II, bioregionalism took on more of a unifunctional character, as the Advisory Commission on Intergovernmental Relations has observed. The states adopted interstate compacts relating to specific issues or needs—water pollution control, flood control, marine fisheries, and forest fire control. Spurred by the Hoover Commission's findings of natural resources overlap and inefficiency, institutional experimentation began in the water resources field. The earlier federal interagency arrangements for studies and coordination were expanded to include state representation, as in the case of the New England–New York Interagency Committee (NENYIAC) and its counterparts throughout the country. The states, in turn, sought to perpetuate their coequal role in water planning through federal-interstate compacts (congressionally sanctioned agreements with both federal agencies and states as signatories). The pioneer venture was the Delaware River Basin Commission, an agency with statutory management functions for this major interstate river basin.

In more recent years, bioregionalism has taken other shapes and directions. A spate of categorical planning programs has arisen, largely under financial incentives and policy direction provided by the Congress. Framework institutions have been created, such as the Title II river basin commissions and the Title V regional commissions. Each of these ventures has been transitory in nature, merely a product of the times and the available federal dollars. They have tended to beg the basic question of how to cope with biological resources on a regional basis in some enduring fashion.

Over the next several chapters, an attempt will be made to trace the history of one such bioregional effort—New England's uncompromising search for a self-determined role in river basin affairs. The story begins with the establishment of the New England–New York Interagency Committee in 1955. It ends with the abolition of the New England River Basins Commission (NERBC) in 1981. It attempts to depict the relationships, attitudes, and interplay among the various participants, and to explain why events did or did not take place.

Like any maturing stream, the course of New England river basins history has meandered widely upon occasion. By mid-century, widespread national interest in river basin management prevailed. Interagency rivalries were beginning to emerge, and the preeminence of the Corps of Engineers in water resources affairs was beginning to be questioned and seriously challenged. The Hoover Commission's sober revelations concerning the low level of efficiency of the national government had spawned considerable public debate and many proposals for reform.

Within New England, a severe economic slump had triggered professional inquiries into the prevailing business and economic climate. These, in turn, had posed thoughtful questions as to whether the region was fully utilizing its water and natural resources potential. In many minds, high power costs were the principal deterrent to economic expansion. However, the prospect of public power projects under federal auspices was stirring New England's latent provincialism as never before.

Despite their previous success in averting various basin authority proposals, New England's business and political leaders were astonished to find a comprehensive resources survey authorized by an obscure item in the Flood Control Act of 1950 and by an unexpected presidential directive. The dual nature of the authorization, and the circumstances surrounding its initiation, augured a particularly controversial beginning for the New England–New York Interagency Committee.

NENYIAC started with the conventional approaches of a federal river basin investigation organized by mutual agreement among the agencies. The states, however, delivered an ultimatum at the outset demanding a measure of direct participation, and an executive council was subsequently formed within which state and federal designees would share equally in the decision-making process. This democratic approach was extended to the deliberations of individual subcommittees and work groups.

It was soon evident, however, that there would be no report unless NENYIAC was viewed as providing an inventory only and not the project authorizing document conventional to such studies. As a result of these developments, a marked change in attitude transpired within the region, with NENYIAC enjoying a measure of credibility and confidence believed impossible five years earlier.

As NENYIAC neared its final stages, the New England governors, at the urging of their designees, leaned toward some vehicle for continued interagency cooperation. The advent of the 1955 floods dramatically underscored this need.

Early in 1956, the Presidential Advisory Committee on Water Resources Policy endorsed the desirability of joint river basin approaches, thereby legitimizing federal participation in such basin coordinating committees. Following numerous charter drafts, the

Northeastern Resources Committee (NRC) came into being in June 1956 by joint authorization of the Federal Interagency Committee on Water Resources and the New England Governors Conference.

NRC's program began with evidence of genuine enthusiasm and many instances of cooperation. A series of state meetings was held during 1958 to acquaint the public with the NENYIAC findings. Gradually, however, the lack of central staff and permanent status led to operating difficulties and uncertainty as to objectives. The state members of NRC finally concluded that a formal federal-interstate compact was the only permanent solution to their problem.

Compact legislation was drafted and introduced into each of the state legislatures and the U.S. Congress. Four of the six New England states adopted the compact, but the opposition of the federal agencies to coequality, plus later opposition from the northern New England senators, spelled disaster for federal ratification despite the favorable precedent set earlier by the Delaware River Basin Compact.

Recognizing the inevitability of the obstacles, NRC transferred its support to the pending national Water Resources Planning Act, which had, by then, been amended to include the coequal provisions so important to New England interests. Detailed consideration of the act's applicability led to the governors' unanimous endorsement of a river basin commission for the New England region in September 1965. For still unexplained reasons, the proposal lay without presidential action until September 6, 1967, when the New England River Basins Commission was officially established.

For the next fourteen years, New England was to enjoy a premier water and related land resources agency. NERBC grew from a chairman and staff of three in 1969 to an agency with nearly fifty employees a decade later. Riding the crest of national environmentalism, the commission left its imprint on virtually every river basin in the region. Not content just to engage in water planning and coordination, NERBC strode into the yet unexplored territory of energy siting and development. Many of its special studies were and still are landmarks in the regional and national planning landscape. It weathered the storm of federal agency politics, reached tentative accord with the states, and left behind a legacy of documents and reports that rivaled the production of NENYIAC. In the proud tradition of its forebear agencies, and in distinct contrast to many of its national counterparts, NERBC was always professional. Of the hardy band of state and federal water resources leaders, regrettably only two of the early spokesmen were around to see their aspirations fulfilled. Time had marched on; the priorities and ground rules had changed nationally; and it was up to the modern crop of agency officials to reinvent the wheel.

Ironically, the federal-state nature of the enterprise, at the heart of the previous NENYIAC and NRC determinations, became the Achilles' heel of NERBC. The failure of its program to address the needs of the states eventually brought about its demise. When the incoming Reagan administration asked the question, "What does the commission do that the states are not able to accomplish themselves?" the answer from the state capitols was a resounding silence.

No representations are made that New England's is *the* model to emulate. In fact, there are those who will assert that much of the period in question was tiresome, inconclusive, even meaningless. Yet there are several persuasive reasons for such a study.

First, the period in question represents a substantial investment in time, money, and commitment on the part of literally thousands of individuals. For these reasons alone, it deserves to be carefully documented before the flow of time erases all previous evidence. Second, with New England again embarked upon a regional water resources organization – the New England–New York Water Council – an analysis of the prior institutional experience may help avoid the mistakes of the past and provide useful clues to the future. And third, the New England river basins story represents an important experiment in bioregionalism. If we are to begin managing natural systems in their entirety as the ecologists would have us do, we must accelerate the search for approaches and institutions that can address resources regionally while remaining credible in conventional political terms. New England's thirty-year span of experience with regional water resources programs represents an important milestone on this long and difficult experimental path.

Acknowledgments

This analysis of New England river basin efforts has been more than a decade in preparation. It began in 1969 with the submission of a doctoral dissertation to Johns Hopkins University entitled *The Thread of the Stream: The New England River Basin Story*, the documentation of the period in New England water resources history extending from 1950 to 1965, which set the foundation for the New England River Basins Commission. In October 1981, the author accepted an assignment from the New England Natural Resources Center, under grant support from the Andrew W. Mellon Foundation, to prepare this sequel — an appraisal of the program of the resultant commission from its initiation in October 1967 to its termination on September 30, 1981. The analysis has special significance because it was written from the perspective of an individual involved directly in New England water resources affairs for much of the three decades in question. The specter of possible bias is more than counterbalanced by the special insights provided by this experience, for many of those consulted were former colleagues.

The procedures utilized in both studies were the same: first, an examination of official reports, documents, and correspondence; then a series of interviews with knowledgeable participants and observers. Thus much of the supporting evidence is primary source material.

It would be impossible to give credit to the many individuals who contributed assistance. At the least, I should mention John E. Sawyer and William Robertson of the Andrew W. Mellon Foundation, who encouraged the project; the trustees of the New England Natural Resources Center, and especially former Vermont Gov. Philip H. Hoff, a New England governor with long-standing regional convictions; the past chairmen of the New England River Basins Commission, R. Frank Gregg and John R. Ehrenfeld, their staff director, Robert D. Brown, and their executive secretary, Rosemary Noonan; and Richard F. Thompson and the staff of the Human Biology Program at Stanford University, who provided logistical and other support during the period of analysis and writing.

I am deeply grateful for the encouragement and assistance received from all quarters, but especially appreciative of the rare opportunity afforded a practicing professional to review and appraise a significant piece of history from beginning to end. Few ever have that privilege.

Needham, Massachusetts   Charles H. W. Foster
June 1983

Abbreviations

CCJP	Comprehensive Coordinated Joint Plan
CIO	Congress of Industrial Organizations
CRSS	Connecticut River Supplemental Study
EPA	U.S. Environmental Protection Agency
FIARBC	Federal Interagency River Basins Committee
FPC	Federal Power Commission
FRC	Federal Regional Council
FWPCA	Federal Water Pollution Control Administration
HUD	U.S. Department of Housing and Urban Development
IACWR	(Federal) Interagency Committee on Water Resources
INCODEL	Interstate Commission on the Delaware
INCOPOT	Interstate Commission on the Potomac
ISC	Interstate Sanitation Commission
LISS	Long Island Sound Study
NAD	North Atlantic Division (Corps of Engineers)
NAR	North Atlantic Region Study
NED	New England Division (Corps of Engineers)
NEGC	New England Governors' Conference
NEIWPCC	New England Interstate Water Pollution Control Commission
NENYIAC	New England–New York Interagency Committee
NENYWC	New England–New York Water Council
NERBC	New England River Basins Commission
NERCOM	New England Regional Commission

NESCAUM	Northeast States for Coordinated Air Use Management
NEWS	Northeast Water Supply Study
NOAA	National Oceanic and Atmospheric Administration
NRC	Northeastern Resources Committee
OCS	Outer Continental Shelf
OMB	(Federal) Office of Management and Budget
ORSANCO	Ohio River Sanitation Commission
RALI	Resource and Land Investigations (U.S. Geological Survey)
SENE	Southeastern New England Study
TVA	Tennessee Valley Authority
WRC	(Federal) Water Resources Council

EXPERIMENTS
in BIOREGIONALISM

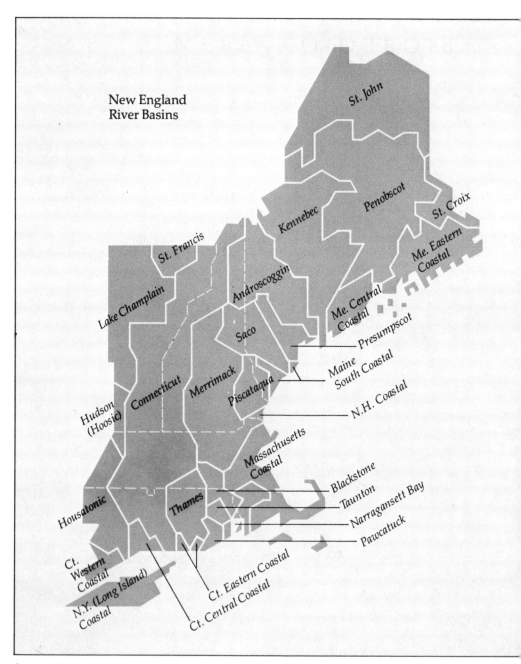

New England
River Basins

St. John

Penobscot

Kennebec

St. Croix

Me. Eastern
Coastal

St. Francis

Androscoggin

Me. Central
Coastal

Lake Champlain

Saco

Presumpscot

Maine
South Coastal

Connecticut

Merrimack

Piscataqua

N.H. Coastal

Hudson
(Hoosic)

Massachusetts
Coastal

Blackstone

Taunton

Housatonic

Thames

Narragansett Bay

Pawcatuck

Ct.
Western
Coastal

Ct. Eastern Coastal

N.Y. (Long Island)
Coastal

Ct. Central Coastal

Source: NERBC

CHAPTER ONE

Introduction

The year 1950 was to be significant for the nation in many respects.[1] Overseas, the United States found itself engaged in its third major military struggle of the half-century. Just before the outbreak of the Korean War came the painful decision to proceed with the hydrogen bomb. Civil defense was fast becoming a household word, and the posture of the country was of fear of wholesale holocaust.

Despite what some observers referred to as a "plethora of prosperity," the nation was plainly on edge. The year began with an assassination attempt on President Harry S. Truman. A finding of perjury was brought against Alger Hiss, as McCarthyism rode the crest of that great new industry and cultural force—television.

In the mid-term congressional elections, misfortune seemed to dog the footsteps of the Fair Deal. Several new faces appeared on the scene. A West Coast unknown named Richard M. Nixon beat veteran Rep. Helen Gahagan Douglas in a particularly vitriolic campaign. Sen. Paul H. Douglas of Illinois was termed "the most promising, most controversial freshman the Senate had seen in years." A freshman senator, Lyndon Baines Johnson of Texas, made the news by outlining a commendable set of ground rules for himself: "Don't spend time looking for headlines; try to avoid politics; be constructive and impartial."

The year 1950 was also to be a year of personalities. The President of the United States addressed some choice expletives to *Washington Post* music critic Paul Hume following Hume's disparaging review of Margaret Truman's first professional concert. Truman, in turn, found himself characterized by presidential aspirant Harold E. Stassen as a "clever politician . . . and . . . the worst President ever to occupy the White House."

Playwright Clare Booth Luce said of Mrs. Franklin D. Roosevelt, "No woman has ever so comforted the distressed—or distressed the comfortable." Women's attention, however, was almost fully preoccupied with that alarming new skirt length, mid-calf.

It was Citation all the way at California's Golden Gate Park for a new world record one-mile race. Bantam Ben Hogan returned from a near fatal automobile accident to win the National Open Golf Championship. A promising star named Carol Channing appeared on Broadway. Comedian Fred Allen described the new toast of television, Arthur Godfrey, as "the man with the barefoot voice."

In Washington, D.C. a welcoming ceremony for visiting Japanese dignitaries was held up because officials could not find the keys to the cabinet that contained the keys to the city.

The *Cheektowaga* (New York) *Times* ran the following classified advertisement: "*Wanted*—man to wash dishes and two waitresses."

And at Brooklyn's Ebbet's Field, the irrepressible Dodgers responded to New York's growing water shortage by drilling not a line drive but a well along the first base line.

As January ushered in the second session of the 81st Congress, it was evident that 1950 was also to be an important year for conservation.[2]

Many of the real issues, however, had been emerging for several years. Two important legislative acts were halfway through Congress, their sponsors rightfully predicting a promising future for the proposed National Science Foundation and the National Trust for Historic Preservation. The 1948 Taft-Barkley Water Pollution Control Act had launched the nation on what was to be many decades of struggle against water pollution. Important implementing programs were to be considered by the second session of Congress. President Harry S. Truman had given fish and wildlife interests an unexpected setback by vetoing the Dingell-Johnson federal aid to fisheries program just as 1949 came to a close. Fortunately, a successful compromise was waiting in the wings for congressional approval. Under the expert guidance of chairman Will Whittington, a $1.114 billion Omnibus River and Harbor and Flood Control Act had been reported by the House Public Works Committee, which had pontificated about "another forward step in the important national programs for development and improvement of rivers and harbors and for protection against the ravages of flood waters."

Despite these impressive accomplishments, the Congress had not entirely lost its sense of proportion. The distinguished Robert A. Taft of Ohio, for example, found the time to take the floor of the Senate in defense of the lowly District of Columbia starling. Due deliberation was also afforded the matter of 750 buffalo, which Interior had deigned to transfer to the Crow Indian tribe of Montana back in 1934 without the express sanction of Congress.

Among the really big conservation issues of the time were those related to the structure of the national government and its administration of water and other natural resources. Water resources planning in general and river basin management in particular had reached critical stages. The events leading to this situation are worth reviewing briefly at this point.

THE NATIONAL SETTING

Some of the earliest traces of our modern consideration of water resources planning can be found in the reports of the Inland Waterways Commission (1908), the National Conservation Commission (1909), and the National Waterways Commission (1912).[3] Although navigation was the aspect of paramount interest at the time, these reports discussed navigation for the first time in the context of other uses, underscoring the need to consider river systems as entire basin units. Coordination of the various federal public service agencies was described as a growing problem even at this early date.

The need for some better definition of national policy relative to hydroelectric power and navigation brought about the enactment of the Federal Water Power Act in 1920, which created the Federal Power Commission (FPC) with the responsibility for evaluating each hydroelectric power project in relation to the comprehensive water resource needs of its basin.

Another important step was the passage of the River and Harbor Act of 1927. Although primarily a relief measure following the major floods of the northeast and midwest sections of the country, the act also embraced the provisions of House Document 308 of the 69th Congress (1st Session). In this document, the Corps of Engineers and FPC were proposing systematic surveys of each major river valley in the interest of flood control, navigation, and power and irrigation development. Congress assigned the Corps of Engineers the task of completing these so-called 308 reports, which, in turn, gave the federal government its first substantial entree into regions such as New England.

In 1933, the Tennessee Valley Authority Act was passed, the first of a long series of innovative acts in the natural and water resources fields. Extensive federal relief efforts came to life under New Deal auspices, including such agencies as the Civil Works Administration, the Public Works Administration, the Works Progress Administration, and the Civilian Conservation Corps. The National Planning Board of the Public Works Administration was to leave a particular mark on New England by making the first federal funds available to the states for planning purposes. A prerequisite, however, was the creation of statutory planning agencies in each of the states.

Establishment of state planning boards led, in turn, to the creation of the New England Regional Planning Commission in the spring of 1934. The commission's Water Resources Committee, working closely with federal relief engineers, began to chart the region's river basin resources systematically, thereby providing New England some of the earliest valleywide, technical reports on water resources in its history.

By 1934, however, the New Deal improvisations in the national water field had become of serious concern to Congress. In response to a specific congressional resolution, President Franklin D. Roosevelt appointed the interagency President's Committee on Water Flow to develop a series of coordinated water development projects for Congress. The report prepared did little more than confirm the lack of agreement and coordination within the executive branch on water resources affairs. In partial response to this situation, the President established the National Resources Board to contrive a better-coordinated program of national and public works planning.

During 1936, devastating floods revisited the New England and Mississippi Valley regions. Congress responded by authorizing another relief program, which included some $10 million in flood control authorizations for the Connecticut River alone—but with substantial strings attached as far as New England was concerned. Although advance consent was provided for the interstate flood control compact on the Connecticut, which the region had sought for years, there was to be for the first time a truly *national* flood control program under the auspices of the Corps of Engineers, wherein actual title to the lands required for reservoir construction would be vested in the United States. This was a substantial departure from previous national policy and a devastating setback to the ardent states' rights attitude of the New England region.

Two years later, the Flood Control Act of 1938 nationalized flood control still further. In sharp contrast to the previous policy requiring local cost sharing, the federal government would now assume all costs of reservoir construction for flood control purposes, including those of lands, easements, and rights-of-way. New England's inability to achieve consensus on flood control had now assured the region exactly what it didn't want—federal development of water resources.

In the meantime, through the efforts of the redesignated National Resources Planning Board, a tri-partite agreement had been hammered out between the Corps of Engineers, the Department of the Interior, and the Department of Agriculture. This device assured at least tenuous forms of coordination on water projects during the late 1930s and early 1940s, an effort that was supplemented still further through Executive Order 9384, which required all projects to be cleared through the Bureau of the Budget.

Following the demise of the National Resources Planning Board in 1934, the Corps of Engineers took the initiative to establish a successor four-party agency, the Federal Interagency River Basins Committee

(FIARBC), adding FPC to the original three agencies. The well-publicized competition of the Corps of Engineers and the Bureau of Reclamation for projects within the Missouri Basin was raising serious doubts as to the credibility of all federal water planning programs.

Resolution of the Missouri Basin issue occurred in the Flood Control Act of 1944, which authorized the first projects under the so-called Pick-Sloan Plan, a compromise program to be carried out by both agencies. The act, however, actually performed double service by quenching a vitriolic conflagration in Vermont's West River Valley over reservoirs authorized for construction in the Flood Control Acts of 1936 and 1938. To Vermont's immense satisfaction, the 1944 act declared it to be national policy henceforth that all completed reports submitted to Congress would include formal comments from the states.

Administrative efforts to improve water resources planning continued to advance during the late 1940s. Under the aegis of the informal FIARBC, now augmented by Commerce, Labor, and the Federal Security Agency, field interagency committees had been established for the Missouri (1945), the Columbia (1946), and the Pacific-Southwest (1948) river basins. These committees served as voluntary coordinating and planning agencies, deriving meager authority both from their filial ties with FIARBC and the working relationships that developed among the various federal and state participants.

With national water policy still essentially undefined, executive agencies flagrantly rampant, controversial proposals for river valley authorities pending, and Congress still determined to play the dominant role in water resources decisions, the stage was well set for the first Commission on the Organization of the Executive Branch of the Government (the Hoover Commission). The commission's majority report, filed with Congress in 1949, called for dramatic changes in national water policy, including a consolidated water development and use service in the Department of the Interior (absorbing the Corps of Engineers), a presidential board of review for water resources projects, and the establishment of advisory organizations within major river basins for improved planning and coordination.

A minority of the commission, including vice-chairman Dean Acheson, had gone so far as to recommend a federal department of natural resources to embrace all the land and water responsibilities of the national government. This drastic step was advocated as the only alternative to an otherwise inevitable proliferation of regional river basin authorities.

Despite the substantial policy recommendations already formulated, a second review of national water resources policy was undertaken in 1950 by the President's Water Resources Policy Commission (the Cooke Commission) established by President Truman. In its report of December 1950, containing more than one hundred specific recommendations, the commission supported the concept of a federal department of natural resources, as well as the earlier Hoover Commission recommendations of a water development service and a board of review for water projects. Also advanced was the suggestion of interagency basin commissions under an independent federal chairman to make the regional or valley authority device less objectionable.

By 1950, however, the public clamor over river valley authorities was great all across the country. A Missouri Basin Authority proposal was before Congress for consideration. The Upper Colorado River Basin Compact had been signed, marking the successful conclusion of nearly three decades of painstaking negotiation between the United States and the water-short states of Arizona, Colorado, New Mexico, Utah, and Wyoming.

Barely a week after signing the Colorado River legislation, President Truman had submitted to Congress a special message recommending the establishment of a Columbia Valley Administration. This new federal agency, tactfully titled an administration rather than an authority, proposed to consolidate within a single agency the functions of the Bureau of Reclamation, the Corps of Engineers, and the Bonneville Power Administration, as they related to multiple-purpose conservation and use of water in this river basin.

The public outcry against valley authorities was not long in coming. Wyoming Gov. A. G. Crane termed the proposed Columbia Valley Administration "an invasion of states' rights." Rep. Harris Ellsworth of Oregon described it as "a proposal to bind most of the five states in the Pacific Northwest in the chains of a Columbia Valley Authority."[4] The influential *Des Moines* (Iowa) *Register* editorialized against river valley authorities in general, and the proposed Missouri Valley authority in particular. In an interview carried by the *Omaha* (Nebraska) *Sunday World-Herald,* Rep. Ben Jensen of Iowa described the program of regional valley authorities as "the recommendation and hope of the Communist Party of America."[5]

The International Association of Game, Fish and Conservation Commissioners, convening in the shadow of the controversial TVA, devoted a substantial share of its Memphis meeting to a pro and con discussion of federal river basin management. Not so the 150,000-member National Wildlife Federation, which adopted a resolution at its annual meeting expressing itself as "unalterably opposed to the creation of any additional federal regional or valley authorities as being unjustified, unnecessary, and a dangerous departure from our American form of government."[6]

By a curious stroke of coincidence, it fell to the lot of federation President David A. Aylward, a lifelong resident of Peabody, Massachusetts, to convey these sentiments to the Congress. The task was not at all distasteful, for much was also stirring within New England on the conservation scene, and the federation's position regarding valley authorities was clearly in tune with that of the region.

THE REGIONAL SETTING

No real sense of urgency had motivated New England until the floods of 1927, which devastated the Connecticut Valley in what was reported as the worst such event in three centuries of recorded history. A decade and a half later, the bulk of Vermont's bonded indebtedness could still be attributed to its flood reconstruction program.

Massachusetts Institute of Technology professor and consulting engineer, H. K. Barrows, was engaged first by Vermont and later by New Hampshire to examine the situation. His solution was a series of multipurpose hydro and flood control structures whose costs would be defrayed by the sale of power. The Boston Society of Civil Engineers, in appraising the 1927 flood, agreed that the development of additional power storage, preferably under private auspices, was the logical approach to New England's flood control needs. This, however, would also require the concurrence and active participation of the water resources agencies of the separate states.[7]

The States

At the time, Vermont, like many of her sister states, had only fragmented functions in the water resources field. These were divided among the public works, public service, and public health agencies. With the establishment of the Vermont State Planning Board in 1933, provision was made for the employment of a professional hydraulic engineer, and for the next fifteen years the public works and planning agencies were to share the spotlight as the most prominent spokesmen for the state's water interests. By 1947, through the combined efforts of Philip Shutler of the Planning Board and civic leader Richard Rogers, the state legislature had been persuaded to establish the Vermont Water Conservation Board. By 1949, this agency had begun exercising a measure of general oversight for water resources.

In New Hampshire, the genesis of water resources functions had taken a somewhat different turn. For many years a small group of business and political leaders, including Laurence Whittemore of the Brown Company, and Gov. Robert Bass and Gov. John Winant, had felt that New Hampshire needed a state water resources agency. At a meeting in Winant's house one Sunday morning, the Barrows recommendations for flood control were discussed, and a decision was made to go ahead with the long-considered water agency. Gov. Styles Bridges, a former Farm Bureau official, was persuaded to endorse the proposal, and in 1937 the New Hampshire Water Resources Board came into being. Whittemore became the board's first chairman.

At the time of the board's establishment, flood control was a priority item in the state. However, attention was also being directed toward the development of New Hampshire's water resources now that its rivers and streams were no longer needed for log drives. Following the 1936 floods, the board's unique structure as a public corporation was used to undertake the construction of the Pittsburg Reservoir on the upper Connecticut River for flood control and power purposes, a project that had been suggested originally by Professor Barrows.

Public interest was also emerging in relation to pollution abatement, and in 1937 the legislature established a separate New Hampshire Water Pollution Control Commission to assume these responsibilities.

New Hampshire's federal agency relations, in contrast with Vermont's, were relatively cordial prior to 1950, and the Corps of Engineers was invited into the state repeatedly to conduct flood appraisal studies.

For sparsely populated Maine, state water resources functions would not command a high priority for many years. By 1947, however, the proposed New England interstate water pollution control compact had directed attention to the deteriorating condition of many of Maine's inland waterways. Of particular concern were such interstate streams as the Androscoggin and Saco rivers. Faced with a request to endorse legislative ratification of the interstate compact, Gov. Edmund S. Muskie chose instead to seek the necessary state machinery for pollution control, arguing that Maine's concurrence with interstate stream classifications would be fruitless if it did not have the authority to carry them out.

By 1949, the Maine Water Improvement Commission had been established by legislative action, but no agency in Maine had as yet been assigned general water resources responsibilities.

In Massachusetts, water resources functions were dominated until mid-century by three agencies. The Department of Public Health exercised general oversight over water resources under its broad health and sanitary powers. Its world-famous Lawrence Experiment Station on the Merrimack River was the leading institution of its kind in the country. In the absence of any central water resources agency, the Department of Public Health undertook many of the ground and surface water investigations authorized by legislative action.

Perhaps the longest tradition of involvement with water resources was that of the Department of Public Works. These functions dated back to the river and harbor functions carried out by boards of lands and harbor commissioners established in colonial times.

The Metropolitan District Commission, the City of Boston's primary water supply agency, was the dominant public agency in this aspect of water resources. A succession of development projects had culminated in the construction of Quabbin Reservoir in central Massachusetts during the late 1930s. Although its primary mission related to the needs of the City of Boston, the commission was rapidly becoming a regional water supply agency for the entire metropolitan area and even portions of central Massachusetts.

Although there was a cordial operating relationship between these three agencies, the state's water resources functions were far from properly coordinated. It would take a combination of events—a new federal act, the Watershed and Flood Prevention Act (Public Law 83–566), and the devastating hurricane floods of 1955 to bring into being the Water Resources Commission Massachusetts would need in the future.

It was fitting that the primary water resources functions in Rhode Island, a coastal state, would relate to its salt water regions. Narragansett Bay and its various tributaries displayed the most prominent water resources problems of earliest times.

As was the case with Massachusetts, the public health and public works agencies shared the responsibility for the state's water resources initially. The Providence Water Supply Board was to become the dominant factor in the water supply field with its extensive Scituate Reservoir project of the 1940s. A decade earlier, however, Gov. Theodore F. Green undertook to abolish many of the administrative commissions established by special legislative act, and Rhode Island's present executive structure began to appear. By 1951, the seven-member Water Resources Commission had been established to study and devise a comprehensive state water policy. This action would lead to the creation of the interagency Water Resources Coordinating Board in 1955 to coordinate, if not actually consolidate, the state's water resources functions.

Connecticut's water resources program was one of the most advanced in New England as the decade of the 1950s approached. As in many of its neighboring states, problems of water supply and water quality control had been, historically, among the first considered. By the late 1920s, the state's water pollution control and sanitary engineering responsibilities were centered in the Connecticut State Water Commission. Connecticut was an active participant in the August

1930 meetings in New York that led to the establishment of the Interstate Sanitation Commission for New York harbor and environs. In 1941, it joined New York and New Jersey as an official member of this early interstate compact agency. Following the 1936 floods, the state legislature acted promptly to create two new agencies, the three-member Flood Control and Water Policy Commission and the five-member Board of Supervision of Dams. The technical staff of the State Water Commission provided the operating and engineering services for these agencies as well. It would not be until 1957, however, and another flood emergency, that the state's principal water resources functions would be consolidated within a single agency, the Connecticut Water Resources Commission.

Interstate Compacts

In addition to the pressure for improved state administration of water resources, the tide of interstate compact action was also strong in New England.[8] Though often criticized for its technocratic character, the compact device was appealing to New England at that time. New England's interest in interstate action appears to have arisen shortly after the Civil War as the various state fish commissioners attempted to devise uniform systems of regulation. Until the 1930s, however, there was little real need for intergovernmental devices of this nature.

In 1935, the problem of competitive wage scales in the West and the South led the New England states to draft an interstate minimum wage compact that was later enacted by New Hampshire, Rhode Island, and Massachusetts. Passage of the Fair Labor Standards Act in 1938, however, obviated the full employment of this device. Early compact approaches were also used for the interstate supervision of parolees and the administration of bridges and other boundary areas in the New England states.

Following the devastating 1936 floods, the governors of Vermont, New Hampshire, Massachusetts, and Connecticut each appointed three members to a special flood commission that traveled extensively throughout the region during the remainder of the year. Secretary of War Woodring's suggestion of an interstate flood control compact made at the March 1937 New England Governors Conference fell on receptive ears, and a compact was later drafted and enacted by each of the state legislatures. Last-minute policy differences, however, blocked the necessary congressional consent, and a formal flood control compact was not approved until 1953. Additional flood control compacts were later enacted without difficulty for the Merrimack (1957) and the Thames (1958) river valleys.

The second major water resources compact effort occurred in the field of water pollution abatement. Before 1947, the New England states had been meeting regularly to discuss joint water standards for interstate streams, but no official action had taken place. Alarmed by the growing federal interest in pollution abatement, the region's Conference of State Sanitary Engineers determined to take the initiative itself. The region's already considerable experience in interstate compact action, plus the precedents at hand in the Ohio River Sanitation Commission (ORSANCO), the Interstate Commission on the Delaware (INCODEL), and the Interstate Commission on the Potomac (INCOPOT), enabled the state directors to persuade a majority of the states and the Congress to accept the New England Interstate Water Pollution Control Compact. This measure provided a means by which the region's interstate and coastal waters would receive continuing water quality classification. By 1950, all but the state of Maine were signatories to the compact.

Success with these efforts led to a further utilization of the compact approach. The disastrous Bar Harbor forest fire of 1947 encouraged New England officials to explore an organized method of mutual assistance. Sen. George Aiken of Vermont, with the support of thirteen New England–New York cosponsors in the Senate, was guiding a proposed Northeastern Forest Fire Protection Compact through Congress during the second session of the 81st Congress.

His colleague, Sen. Theodore Green of Rhode Island, with equally impressive cosponsorship, had filed legislation amending the Atlantic States Marine Fisheries Compact to aid in the promotion and development of the coastal fishery resources. The need for improved coordination within the states was underscored by Senate ratification of various international agreements, including one for the northwest Atlantic fisheries, which embraced the traditional fishing banks off the New England coast.

Economic Studies

During the years immediately prior to 1950, the New England Governors Conference was attempting to grapple with a number of serious economic problems.[9] The floods of 1944 and 1946 had renewed the concern for more adequate flood control within the region, and the governors had created a special New England Interstate Committee on Flood Control to seek out practical and equitable solutions.

The brief but intense business recession of 1949–50, plus the pressing demands of the Korean War, had focused concerned attention on New England's lagging economy. An apparent New England economic decline was clearly worrying business and political leaders. In partial response, a New England Development Authority compact had been proposed in 1949 by unanimous action of the governors, but this legislation had fared poorly in most of the state legislatures. A special conference was also held in Washington early in 1951 to boost New England's potential role in defense activities among congressional and government officials.

At the suggestion of chairman Leon H. Keyserling of the President's Council of Economic Advisers, a series of detailed regional economic studies had been launched with New England as the second area considered. A subsequent and more detailed economic study had been undertaken by a special Committee of New England organized by the National Planning Association, an independent, nonpolitical, nonprofit organization headquartered in Washington, D.C. The incentive in this case had come from the Joint Committee on the Economic Report of the United States Congress. Two New England members of this committee, Sen. Ralph E. Flanders of Vermont and Rep. Christian A. Herter of Massachusetts, were sufficiently impressed by the association's just-published report on the South to prevail upon chairman George O'Mahoney to request a similar study for the Northeast.

The results of the economic studies were generally reassuring. However, both reports emphasized New England's need to change its traditional business patterns and, in particular, to make fuller use of its water and land resources potential.

Hydroelectric Power

The most pressing and by far the most controversial problems, however, related to hydroelectric power.[10] Prospects of cheaper electric power through the proposed St. Lawrence Seaway and Power Development Project were enticing the northern New England states away from their traditional private utility orientation to the intense discomfort of the New England Council, which had officially opposed the St. Lawrence Project since 1941. Speculation also prevailed concerning possible chemical and aluminum industry relocations, which would provide a valuable boost to New England's lagging economy—but only if something could be done about the region's traditionally high power rates.

In 1948, the Power Survey Committee of the New England Council, utilizing the consulting services of Charles T. Main, Inc., had struck back with a report estimating New England's undeveloped hydroelectric reserves at no more than 500,000 kilowatts under present economic conditions. This figure was in sharp contrast to the more than 3 million kilowatts assigned to the region by FPC's power market surveys. The discrepancy in estimates and the possible effect of additional hydropower on New England's future economic growth and development were lively topics within the region. Sen. Leverett Saltonstall of Massachusetts, in determined pursuit of the correct facts, had demanded a documented report from FPC, which he received on February 28, 1949. While confirming the earlier estimates of 3 million-plus potential kilowatts for New England, FPC chairman Nelson Lee Smith was careful to state that these were the results of preliminary examinations only.

Public power advocates, however, had been quick to take advantage of even the preliminary figures. Former FPC chairman and Interior official Leland Olds, addressing a forum sponsored by the Hampshire County (Massachusetts) Chapter of Americans for Democratic Action, argued for a Connecticut River Valley Authority to capitalize upon the region's power development possibilities. Assistant Secretary of the Interior C. Girard Davidson somewhat expansively told a Worcester assembly of the Textile Workers Union of America that 3⅓ million kilowatts were available in New England "all alone and without assistance from steam plants." This was an unusual statement to make, since hydro normally required at least three times as much base load power to be operationally feasible.

A legislative recess commission in Massachusetts, spurred on by the published reports of undeveloped hydroelectric power within the state, took expert testimony in October 1949 from federal and state witnesses. FPC reports notwithstanding, the commission's conclusion was that virtually all the feasible sites in Massachusetts had already been developed.

Sensing a lively news story, the *Providence Journal-Bulletin*, late in 1949, asked its financial editor, George H. Arris, to find out which side was right. Arris talked to officialdom at considerable length, then returned to his managing editor convinced that the only way to determine New England's real hydroelectric potential was to take a team of engineers into each major river valley. Three months and twenty-one river basins later, Arris reported finding slightly over 1 million kilowatts as "engineeringly practical," one-third of which lay in the state of Maine. The Maine resources were at that time prohibited from export by state statute, the so-called Fernald Act, an aftermath of Samuel Insull's spectacular but unsuccessful raid on the New England utilities during the late 1920s. The series of ten articles by Arris was reprinted and distributed widely by the New England light and power companies, thereby giving credence to the charge of whitewash made by public power advocates.

The simmering controversy over public versus private power did, however, make New England uncomfortably aware of the potential applicability of the valley authority approach to its own great river basins. The principal threat appeared to lie in the halls of Congress.

The Region and the Congress[11]

Congressional concern for water resources was virtually as old as the nation itself. A measure in support of lighthouses, beacons, buoys, and public piers—probably the earliest river and harbor bill in U.S. history—received the approval of the first Congress in 1789 with what we would now regard as astonishing equanimity. Direct involvement by Congress in water resources affairs derived from the early days of Congress, when it became the practice to refer legislative proposals to special ad hoc committees for detailed drafting before final consideration. The Committee on Interstate and Foreign Commerce, for example, the third standing committee created by the House, included within its jurisdiction all matters pertaining to navigation. The extent of this legislative influence can be illustrated by the fact that from 1946 on, twenty-two of the thirty-four standing committees of Congress were involved in one way or another with water resources.

By the opening of the 81st Congress, New England could consider itself relatively well off in terms of potential legislative influence. With 12 out of 96 votes in the Senate, as compared to 28 out of 435 in the House, its voice on the senior side was, of course, the more substantial.

At the start of the session, Leverett Saltonstall of Massachusetts was serving as minority whip in the Senate, and George D. Aiken of Vermont was the ranking minority member of the Senate Agriculture and Forestry Committee. Their two New Hampshire colleagues, Styles Bridges and Charles W. Tobey, were the second and third ranking Republicans in the Senate in terms of service. Bridges served as the ranking minority member of the Appropriations and Armed Services Committees, while Tobey could have his choice as the ranking minority member of the Banking and Currency, Interstate and Foreign Commerce, Crime Investigation, or Small Business committees.

Maine's Owen Brewster, as chairman of the Senate Republican Campaign Committee to coordinate state and national campaigns and assist Republican candidates nationally, was in a particularly strategic position of influence.

Although New England's Democratic senatorial delegation was slim in number and length of service, its position as representative of the majority party would certainly be of considerable help. Eighty-three-year-old Sen. Theodore F. Green of Rhode Island, a former governor of that state, was regarded as the dean of the Senate and, therefore, a voice to be respected at all times. This reputation was considerably enhanced by his second ranking majority spot on the Committee on Rules and Administration.

On the House side, New England could boast an almost complete sweep of leadership with John W. McCormack and Joseph W. Martin, Jr., both of Massachusetts, serving as majority and minority leaders respectively. Edith Nourse Rogers of Massachusetts, the fifth ranking Republican in the House in terms of years of service, occupied the only key committee post, serving as ranking minority member of the House Veterans Affairs Committee. With so little representation in the House, New England's leadership influence, and not its delegation, would have to be its principal weapon.

Despite this promising potential, attempts at forging a united New England delegation had been casual at best, as compared with the solid front exhibited by many of the western and southern delegations.

During a postcampaign swing through Massachusetts in 1953, newly elected Sen. John F. Kennedy promised to give special attention to the problems of the New England region. True to his word, Senator Kennedy took the leadership in organizing a New England Senators Conference, which met biweekly, and his administrative assistant, Theodore Sorenson, became its first secretary. The conference continued as a cohesive unit under the sponsorship of Sen. Leverett Saltonstall of Massachusetts when Kennedy was elected to the presidency, and became a reasonably effective instrument for advancing the region's interests on the senior side of Congress.

The New England delegation in the House, however, was far less organized. Rep. John W. McCormack of Massachusetts, an individual with strong regional leanings, occasionally exercised his majority leadership influence by bringing together the New England congressmen for conferences on New England problems. As House Speaker, Joseph W. Martin, Jr. sponsored luncheons for the Massachusetts delegation on a regular basis, to which House members from the other New England states were often invited. These events, however, were generally unstructured and characterized by heavy doses of social interchange and political commiseration.

As one looks over the other members of the New England delegation of the time, the privilege of hindsight could identify many who would be heard from in later years. Sen. Henry Cabot Lodge, Jr. of Massachusetts, for example, was already known as an individual unafraid to speak out on what he thought was right. Norris Cotton of New Hampshire was beginning his second term in the House in a political career that was to later lead him to a seat as the Granite State's senior senator. John E. Fogarty of Rhode Island was destined to become an influential spokesman for New England rivers and harbors projects and a leader in national health legislation. And the Massachusetts delegation, with Christian A. Herter and Foster Furcolo, two governors-to-be, and an obscure but promising young second-term congressman, John F. Kennedy, were to leave a considerable imprint on posterity.

Thus, despite the key positions held in Congress, the New England delegation was one more in name than in fact. The business and economic leadership centered in the New England Council appeared to view natural resources more as an underutilized raw material for industry than as an area around which a major regional program could be built. The New England Governors Conference, still little more than a loose alliance of state political leaders, gave promise of becoming a significant regional force but, at best, could cast but a shadow of influence on the national scene. With respect to the state water resources agencies, most were hardly out of swaddling clothes, and even those with considerable longevity enjoyed little public stature or support.

From this brief review, it is evident that New England was more of a region physically and economically than it was politically as the year turned the corner into 1950. Leadership was present, and surprisingly good leadership at that, but New England attitudes were still largely introspective and self-preoccupied.

CHAPTER TWO

The New England–New York Interagency Committee

1950

As the 1950 session began, it was apparent that New England's river basin fortunes were to rest in substantial part upon the actions of Congress.[12] For example, Rep. John E. Rankin of Mississippi had filed legislation authorizing multipurpose federal authorities for each of the eleven principal river systems from the Atlantic to the Pacific. New England's Connecticut River was to be one of them.

Rep. Thomas J. Lane of Massachusetts was the author of specific legislation for a Merrimack Valley Authority, echoing a suggestion first appearing in print a decade and a half previously in the *Boston Transcript*. The Tennessee Valley Authority (TVA) was to be the model for the Merrimack.

Rep. Chase Going Woodhouse of Connecticut was pressing for similar action on the Connecticut River. Her remarks before the National Emergency Conference on Resources on May 13, 1949 were heavily influenced by the fate of the New England Development Authority proposal, endorsed unanimously by the six New England governors but rejected unequivocally in Maine and Vermont earlier in the year. The survey of natural and economic resources envisioned by that interstate compact agency must now be carried out under federal auspices, she argued.

Rep. Foster Furcolo of Massachusetts, purportedly at constituent request, had introduced a third Connecticut Valley Authority proposal. This measure called for unified water control, resource development and generation of low-cost hydroelectric power under a presidentially appointed three-man authority, which was required to meet in the region only twice annually. As with many of his New England colleagues, however, Furcolo's primary interest was in having the region's resources properly examined. There were many such survey proposals before the final session of the 81st Congress.

Representatives Furcolo and Lane had filed study proposals for the Connecticut and Merrimack river valleys in addition to their authority legislation. Lane's bill for the Merrimack River valley was directed specifically toward the question of untapped hydroelectric resources. Their Massachusetts colleague, Rep. Edith Nourse Rogers, was proposing a comprehensive investigation of all New England rivers under federal auspices.

On the Senate side, even Henry Cabot Lodge of Massachusetts, an arch-opponent of the St. Lawrence Seaway in the 80th Congress, was proposing a comprehensive survey of "hydroelectric power, flood control and other improvement on the Merrimack and Connecticut rivers and such other rivers in the New England states where improvements are feasible."[13] The substance of this bill was to be incorporated in the NENYIAC congressional authorization later on.

Sen. Theodore F. Green of Rhode Island had gathered impressive cosponsorship from northern New England and New York for a proposed New England–New York Resources Survey Commission. The Green bill, however, had raised two troublesome points: the inclusion of New York, and the recommendation for a study commission under an independent chairman.

Radical though these measures appeared to be, even more far-reaching proposals were being circulated publicly within the region. In a *New Leader* article on March 12, 1949, William E. Leuchtenburg, Boston director of Americans for Democratic Action, had spoken confidently of the first national conference on river valley development to be held in Washington April 21–23, 1949. "The ultimate solution for New England is an overall regional authority," Leuchtenburg had declared.[14]

The administration, in the meantime, also had its own ideas as to what the Northeast needed. President Truman, in his State of the Union message, had come out foursquare for the proposed St. Lawrence development, and high administration officials were actively promoting the concept of an extensive northeast power grid in public appearances throughout the region.

Despite active administration support, the Green bill had not met with appreciable favor in Congress. In fact, as the result of astute legislative maneuvering by Senators Lodge and Saltonstall and Representative Furcolo of Massachusetts, the Omnibus River and Harbor and Flood Control measure pending in the Senate had been amended to include language in its Section 205 containing the substance of the Lodge proposal. With a suitable survey authorization already proposed for the Corps of Engineers, the Green commission would not be necessary, it was argued.

In a special message to the Senate on February 9, 1950, President Truman underscored the shortcomings of a single-agency study and stressed the logic of including New York in any regional investigation, particularly one in which resources were to be prominently featured. He urged the Senate to modify the omnibus bill further to provide for a special commission survey for New England similar to that for the Arkansas-White-Red region. During Senate floor action the administration amendment, sponsored by Sen. William Benton of Connecticut and others, appeared headed for certain rejection and was, therefore, withdrawn by its sponsors.

As it reached President Truman's desk, the River and Harbor and Flood Control Act of 1950, like all omnibus bills, was a mirror of the needs and desires of individual congressional members. Included among its pork-barrel provisions were some $100 million in new projects never cleared by the administration. Not unlike many of his predecessors, President Truman was clearly on the spot. On the one hand he could sense evidence of declining Fair Deal support in Congress (to be confirmed later by the disastrous November elections). This argued strongly for a policy of not antagonizing what little administration support remained. On the other hand, the omnibus bill contained much that was outright objectionable to the administration, as evidenced by a strong memorandum from the Department of the Interior recommending that the entire measure be vetoed. Among the items singled out for special castigation was the proposed survey of New England river basins inserted at the last moment by Massachusetts interests.

There was, however, no realistic alternative at this point. President Truman, therefore, reluctantly signed the bill but dispatched an accompanying message to Congress on May 22, 1950 pointing out, among other matters, the failure of the measure to provide for a truly comprehensive investigation of the Northeast. He also placed Congress on notice that he would "issue instructions to the appropriate Federal agencies to work with the States in preparing as much of a combined resource development plan for this area as existing law will permit."[15]

The President was not alone in these convictions, for a coalition of five New England senators, plus the influential House majority leader John W. McCormack, had written him urging that he take action under his executive powers.

On October 9, 1950 President Truman dispatched instructions to the six agencies then represented on the Federal Interagency River Basins Committee (FIARBC) requesting the immediate establishment of a special committee to carry out the special survey of New England rivers authorized by Section 205 of the Omnibus River and Harbor and Flood Control Act of 1950.

In simultaneous action, the President also wrote to the seven northeastern governors inviting each to designate a representative to the interagency committee. The key sections of the executive communication read as follows:

The comprehensive study of land and water resources of this area should include, among other matters, coverage of electric power generation and transmission, forest management, fish and wildlife conservation, flood control, mineral development, municipal and industrial water supply, navigation, pollution control, recreation, and soil conservation. The necessary first step in such a study is an inventory of the land, water, and all of the related natural resources available for utilization, together with a survey of the projected regional and national requirements which might be met through more effective utilization of the natural resources of the region. When these basic facts on resources and needs have been collected and analyzed, the committee should then proceed to determine what development and conservation projects are feasible and desirable, and to prepare recommendations for specific action to carry them out.[16]

Meeting in special session on October 27, FIARBC moved with dispatch to authorize a New England–New York Interagency Committee for the Northeast under the chairmanship of the Corps of Engineers to carry out the presidential mandate. Mindful of its posture as consultant to Congress, the Corps reserved to itself the option of submitting a separate report to Congress under Section 205 of the Flood Control Act, taking the strategic stance that the executive and legislative directives were related but separate actions.

With New York also included in the study, the Corps had to decide whether the New England Division (NED) or the North Atlantic Division (NAD) would be its primary administrative unit. NAD won out after the Federal Power Commission (FPC) questioned the cordial relationship between NED and the New England utilities, arguing that public power might not receive a fair shake under NED auspices.

At an agency summit meeting sponsored by NED in New York, NENYIAC's committee structure was laid out and subcommittee chairmanships assigned to the appropriate federal agencies. These actions appear to have been accomplished with relative equanimity. The interagency wranglings of the Missouri Basin studies were still fresh in bureaucratic minds. Consequently, the subcommittee pattern was designed to ensure maximum participation and discussion by all agencies. The fact that the assignments were made in advance of any formal organizational meeting and without prior consultation with the governors' representatives tended to confirm the prevalent feeling throughout New England that NENYIAC was just another government study.

Although most of the federal agencies elected to utilize their regular administrative channels, Interior decided that a separate NENYIAC office in Boston would be warranted. Mindful of the public power implications, which had been so instrumental in NENYIAC's creation, former FPC chairman Leland Olds, then director of Interior's Division of Water and Power, was given the nod as official NENYIAC representative. To make the assignment somewhat less blatant, Olds and his assistant, Mark Abelson, were subsequently transferred to the resources program staff, which functioned from then on as the official spokesman for Interior.

By the time of the February FIARBC meeting, it could be reported that all of the federal representatives had been named, and that the majority of the governors had made their official designations.

Meanwhile, New England's political leaders, congratulating themselves on having maneuvered various valley authority and resource survey bills into oblivion, had awakened to find their labors circumvented by direct presidential action. Although NENYIAC was merely a survey authorization, its antecedent manipulations, plus the clearly federal dominance, raised suspicions that the investigation was nothing more than an opening wedge for unwanted federal programs.

Connecticut's Gov. John D. Lodge complained publicly about the money now earmarked for NENYIAC, which should have been going to higher-priority New England projects. Editorial doubts were expressed by the *Hartford Times* which termed NENYIAC merely a "facade" for more grandiose federal schemes.[17]

Gov. Thomas E. Dewey of New York expressed his reservations by appointing chairman John E. Burton of the New York Power Authority as his state's official representative. As Burton told NENYIAC at its first meeting: "I must say we are highly suspicious that there is a move to federalize the natural resources of all the States."[18]

Vermont's representative, Philip Shutler, spoke publicly of his state's attitude of "watchful cooperation."[19] In private, however, he coined the derisive term *knickknack* to characterize NENYIAC's seemingly all-encompassing objectives.[20]

Gov. Sherman Adams echoed the general feeling of skepticism in New Hampshire when he said: "The conclusions of this committee . . . must not be based on somebody's preconceived notions of what is best for the country as opposed to what is best for the people of New Hampshire."[21] The general attitude was best illustrated by a speech presented by the state's NENYIAC representative, chairman Walter G. White of the Water Resources Board, before a state dinner of the New England Council, entitled "A Little Bureaucrat Looks at a Big Bureaucrat."[22]

"The big bureaucrats want to study us as though we were a backward child," White stated, alluding darkly to the

sinister forces at work within and on the fringe of this Committee (NENYIAC). . . .

If the trend of the big bureaucrat is not arrested, we will find ourselves in the status of a seedy moocher living on handouts from Washington. NENYIAC is another milepost along the road.

NENYIAC, however, was not entirely bereft of support. Gov. Paul A. Dever of Massachusetts told the first meeting that he was "suspicious of those who fear the facts." He urged the committee to conduct a factual study without preconceptions, stating, "Let no special interests stand in the way of a study which at the very worst can do no harm, and which at the best can be a forerunner of new vistas of beauty and areas of recreation in our river valleys. . . ."[23]

With these faint words of encouragement NENYIAC set about its appointed task. A schedule of meetings was agreed upon that would bring the committee to each of the state capitals over the next seven months. These would be followed by an initial series of public hearings, one in each state, to ascertain topics of particular concern within the region. The investigative phases were officially broken down into eleven major study and report groups with three additional subcommittees (mapping, hydrology, and economics) to provide across-the-board data for all work groups. The federal agency personnel plus one designee from each state were invited to membership on each subcommittee and report group.

Vital questions soon arose as to what standards were to be used in accepting or rejecting potential projects and how final decisions were to be made. It was here that traditional Yankee individuality rang strongly in the Army's ear. Col. F. F. Frech, NAD engineer and chairman of NENYIAC, outlined the Army's policy as follows: "We want — and will exert every effort to maintain — the town meeting approach to both the studies undertaken and to the problems they evoke."[24] Despite such virtuous statements, it was evident that all was not well with NENYIAC. It was at best an uneasy truce that prevailed among the federal designees and their state counterparts.

1951

Early in 1951, Connecticut's representative told the *Hartford Courant* that NENYIAC had agreed to postpone the question of administrative jurisdiction over natural resources until the studies had been completed. This was, he said, a clear victory for the states, which were vigorously opposing the federal agencies "grabbing power" over resources.[25]

It was appropriately in Hartford, Connecticut, site of the revolutionary Hartford Conventions, that the first real confrontation took place. The state representatives delivered an ultimatum to chairman Frech. They wished it clearly understood that the states would not participate merely as advisers but demanded an actual voice in all decisions.

When Col. Benjamin B. Talley assumed the NENYIAC chairmanship in May 1952, it was evident that his prior experience in military intelligence would stand him in good stead. Blond in appearance and forthright in manner, he faced the incipient revolution characteristically head-on.

The situation was compounded by NENYIAC's pilot project on the Androscoggin, which had ground to a halt on the issue of whether power project benefit-cost ratios would employ government or private rates of financing. The issue was a critical one for public power advocates. Not only were actual project authorizations at stake, but major blocks of the FPC-identified New England hydro were on streams such as the Androscoggin.

At a session in Boston chairman Talley told NENYIAC that he had no specific instructions on state participation. His only directive from the Office of the Chief of Engineers was to complete the report as authorized and on schedule. He suggested the formation of an executive council in which each state and each federal agency would be entitled to one seat and one vote. All nonconsensus decisions would thereupon be determined by simple majority vote.

Although chairman Talley's proposal was the only practical course at this point and was substantiated to a degree by the state relations directive given the Corps in the Flood Control Act of 1944, the move was regarded as a milestone by states' rights advocates. It received swift approval. From then on, NENYIAC itself virtually ceased to exist, its functions carried out pro forma by the executive council.

The policy of decision by majority vote carried forward into subcommittee and work group deliberations. It marked the beginnings of real state participation and, eventually, real enthusiasm for NENYIAC on the part of its state members. It enunciated the principle of coequal representation later confirmed as national policy in the Delaware Basin Compact legislation and the Water Resources Planning Act of 1965.

With its administrative details now firmly in hand, NENYIAC proceeded with the initial round of public hearings. They were to all intents and purposes a near disaster. The prominent issues of public power, valley authorities, and large versus small dams were quick to appear.

The first public hearing was held in Springfield, Massachusetts, on June 27. The opening witness was Anthony Wayne Smith, architect of the post-war employment program of the Congress of Industrial Organizations (CIO), whose interest in water resources and regional planning derived from its continuing efforts to improve the social conditions of the working man. Smith's personal recollections of conditions in the Pennsylvania coalfields made him an understandably strong proponent of the "cleaner" hydro power approach. Smith made it abundantly plain that the CIO was interested in low-cost power for the region. He also took a dim view of the federal dam construction program, except for the generation of power, which he termed "the first purpose of dam construction."[26] A combination of power dams plus upstream soil conservation and reforestation, plus local flood control works, should do the job of flood control adequately within any river basin. He even spoke favorably of a TVA-type agency for the Connecticut Valley. The CIO had done its homework well, for Smith was supported by various witnesses from agricultural, rural cooperative, and soil conservation district organizations who stressed the need for a fundamentally upstream approach to flood control problems.

Rebuttal came afterward in the form of testimony from Howard Turner, Harvard Professor of the Practice of Engineering and the principal consulting engineer for the Electric Coordinating Council of New England. Initiated during the wartime years, the council had emerged as a promising instrument for policy and program coordination among New England's scattered and highly independent private utilities. Turner argued that the estimates of potential power prepared by FPC for New England were grossly inaccurate. In fact, he said, the region could absorb economically only about 750,000 kilowatts of peak power in the next 20 years. The FPC's 3 million kilowatt water power figure, even if correct, would require some 9 million kilowatts of base power to back it up. This could not possibly be provided in the foreseeable future by either steam generation or any surplus power from the St. Lawrence River development that might be available to New England.

Turner was subjected to close questioning by NENYIAC members. During the interchange it became evident that the disparity in figures was primarily the result of varying approaches — FPC computing long-range potential from raw data at hand, and the industry considering only projected market demand under current economic feasibility standards.

Ruth N. Gilmore, of the Natural Resources Workshop of Rhode Island, then expressed what was to become a familiar theme of conservation groups throughout the hearings: that the general public believed NENYIAC's primary mission to be the development of federal hydroelectric power projects throughout the region.

The public hearing in New York State was held at Syracuse on October 16.[27] There was even less sympathy for the NENYIAC investigation.

State Assemblyman Wilmer Milmoe expressed the legislature's fear that NENYIAC would somehow lead to a huge central agency governing all the resources of the region. Introduced into the record was a resolution adopted at the Annual Executive Conference of the New York State Committee on Interstate Cooperation the month before, stating, "We support the inherent rights of States and local governments to regulate and utilize their own resources." A. B. Racknagel of the St. Regis Paper Company, representing the New York State Forest Industries Committee, said that "the tradition of New York State is perhaps best exemplified by the phrase, 'We paddle our own canoe.' . . ." He made it plain that the paddling should occur "without any Federal help or hindrance."

1952

The official Vermont hearing convened at Montpelier on March 19,[28] with some one hundred persons on hand for the occasion. The session began mildly enough with tempered remarks concerning the value of upstream agricultural practices in relation to the control of stream-bank erosion and the validity of the agricultural conservation grant-in-aid program.

General manager Thomas Farwell of the Ryegate Paper Company spoke for many when he observed, "I don't know just what you are trying to do." Others, however, seemed more certain. President Albert Cree of the Central Vermont Public Service Corporation accused federal public power of being unsound, socialistic, unneeded, and unwanted. Representatives of Interior and Agriculture rose to defend the federal position, and the altercation grew so heated that chairman Frech was finally forced to intercede.

Clarence W. Mayott and Howard M. Turner of the Electric Coordinating Council argued that the amount of economically feasible undeveloped power was not as important as the effects on the cost of electricity and the power pattern as a whole. Turner presented figures indicating that the net reduction in rates at maximum would be only in the order of two and one-half percent if all of the presently identified sites were to be developed. Manager Walter N. Cook of the Vermont Electric Cooperative, Inc. came right back with an accusation that the private utilities were squandering millions of the consumers' money on advertising campaigns against public power, trying to frighten the public into confusion by half-truths and distortions.

Although several witnesses claimed that Vermonters were not obstructionists by nature, NENYIAC must have had cause to wonder as the day-long hearing concluded. Chairman Frech spoke tactfully of "the spirit of America where you can get up and say what you think," adding, "I feel that Vermont and New England may be better off for having had such a meeting."

At the Maine meeting, held in Augusta on June 12,[29] a delegation from Washington County appeared in support of the Passamaquoddy Tidal Power Project, which State Rep. Spencer Gay described as "this undeveloped pocket book of this earth." The proposal elicited close questioning from the committee on whether there was a demonstrated need for the additional power in Maine. Again, the private utilities made their opposition to federal power projects clear.

As if NENYIAC did not have enough troubles, another controversial subject arose—that of water pollution. Edward W. Atwood of the Bates Manufacturing Company and J. Elliott Hale of the Maine Water Improvement Commission felt that any problems in this respect were already being taken care of. Leonard W. Trager, the Federal Security Agency's NENYIAC representative, took strong exception to these remarks. A native of New Hampshire, Trager had conceived and headed the state's first Water Pollution Control Agency and knew whereof he spoke. His strong convictions and uncompromising manner, however, tended to bring all of New England's native stubbornness to the fore, and he was destined to leave NENYIAC in later years convinced that all hands were against his agency's program.

The predominant sense of the hearing was well summarized in the prepared statements submitted by president Earl B. Foss of the Wilton Woolen Company. "Let us citizens of Maine keep the Federal Government out of this State as much as possible and help to get Federal Government out of business, and let private enterprise function as it used to before the New Deal era."

Misunderstandings as to the basic objectives of NENYIAC continued to flow. A telegram from the Massachusetts Farm Bureau Federation read, "Aggressively opposed to creation Connecticut Valley Authority with Federal control of water and power."

Concord, New Hampshire, on October 9,[30] was not much different. Rural interests were present in force. They were backed up by a joint resolution of the state legislature, presented by House Resources, Recreation and Development Committee spokesman Katherine Jackson, urging the inclusion of forest management and soil conservation practices in all federal flood control programs within the state.

President Avery R. Schiller of the Public Service Company of New Hampshire advised the committee that no industry had ever been lost in New Hampshire because of power rates, and the New England Council offered a reward of $100 to anyone providing specific information to the contrary.[31]

NENYIAC also heard from a newly formed organization, the Connecticut River Watershed Council, destined to become a significant and responsible figure within New England in the years to come. Formation of the council as a regional private action entity was rumored to be the utilities' alternative to the much discussed Connecticut Valley authority.

The frosty reception afforded NENYIAC thawed somewhat during the remainder of the public hearings. Governor Lodge of Connecticut opened the session in Hartford on September 11,[32] by repeating a statement he had made earlier to NENYIAC in executive session.

We, in New England, frequently are accused of carrying to extreme on occasion our traditional independence of spirit. That is perhaps true, yet it is this very attribute which has enabled us, despite our relatively meager natural resources, to build here a prosperous economy and a civilization of inspiring and enduring worth.

In Providence, Rhode Island, on November 13,[33] only a small crowd was on hand to greet NENYIAC. The sentiments expressed were familiar anyway: concern about the steady encroachment of the federal government and about possible dominance of the study by power and flood control interests.

Discouraging though the hearings were, they did provide spice and flavor in a form only New England could supply.

Farmer Hugh C. Tuttle of Dover, New Hampshire, proudly traced his tenure ten generations back to a land grant from the king of England.

Washington County (Maine) haberdasher and Passamaquoddy advocate, Arthur Unobskey, was told that his assertions of unsatisfied power demands in Maine were as artificial as a shortage of ten-dollar suits.

Throughout the hearings, periodic bulletins were received from a flooded property owner in Beacon, New York, who signed herself, "a desperate Mother and more than 24 Taxpayers and their 24 small children."

A self-styled "typical private landowner" in Maine confessed to the fact that he had been asked to testify as such by the Associated Industries of Maine. Worse still, he turned out to be an attorney who represented many of the companies.

But it was City Manager Richard Martin of Manchester, Connecticut, who really topped the list. Martin remarked that the hearings reminded him of a report he had once written for a governor of Connecticut that was mistyped to read, "We had dissipated in a number of conferences!"[34]

Meanwhile, NENYIAC's various task forces and subcommittees had already begun work, and a coordinating center was established in Boston, with staff who were on loan from NED.

What to call the new interagency committee created a brief flurry of excitement. There is some evidence that the New England Division personnel assigned to the Boston coordinating office quietly placed New England ahead of New York in the title, thereby creating NENYIAC instead of NYNEIAC. New York clearly did not like being second-best to New England in this or any other enterprise. The move may also have been a silent protest by NED to the selection of NAD as the central administrative entity for the study, while NED did what it considered to be the bulk of the work.

Following a brief regime under a "spit and polish" officer of the old school, the difficult task of NENYIAC coordinator was undertaken by Col. Harry Fox, described by his associates as a born diplomat. Previously a post engineer with some TVA experience but basically no other background in planning, Fox operated astutely and effectively, drawing out the principals in any argument until a satisfactory solution could be reached.

His successor, Col. Gerard B. Troland, fresh from military service with the Mississippi River Commission, brought new dimensions of precision and organization to the NENYIAC effort at a time when they were most needed. Troland recalled that Fox, on the eve of his retirement, brought a pile of documents into Troland's office, stating, "I hereby turn over to you the greatest batch of unfinished work you have ever seen!"[35]

The military precision of coordinator Troland then became very useful. A color-coded flow chart was maintained on the office walls at 150 Causeway Street, Boston. By sheer coincidence the number of steps required to complete the report was exactly 308, the number traditional to the Corps from its earlier Section 308 basin studies.

The coordinator's task was infinitely more complex than that of a mere conduit. As preliminary reports came in from the various work groups, they were reviewed for accuracy, editorial content, and compliance with NENYIAC policies. It was then the coordinator's responsibility to see that the final studies were compiled into a comprehensive basin chapter. The coordinating staff would add new sections descriptive of the region and also a final coordinated basin plan.

Rejuvenated in a bright blue cover, the integrated chapter was shipped to the various agencies for review and comment. Their responses were then packaged into a two- or three-page digest and resubmitted to the official members of NENYIAC, the chairmen of the three subcommittees and the coordinators of the eleven study and report groups.

Following receipt of comments, the revised "blue books," now rebound in burnished gold, were reprocessed and furnished to the members of the executive council at least thirty days prior to each scheduled meeting. The unit in question was then discussed and formally approved by majority vote action.

From May 13, 1954, when the first "gold book" on the Androscoggin Basin reached the executive council, until March 10, 1955, when the unit on "Reference Data and Special Subjects (Regional)" was finally approved, the council did little else but review reports. The pace was so hectic, one participant observed, that insufficient time was available for properly editing and consolidating the study and report group materials.

The problems faced by an interagency effort of this sort are always legion, and NENYIAC's were no exception. On the federal side, Congress repeatedly delayed action on key appropriations until well into the new fiscal year, thereby cutting short precious periods available for field investigation. Despite the vigorous efforts of Rep. John F. Kennedy of Massachusetts and his associates in the New England delegation, NENYIAC was also consistently underfunded. This eventually forced an extension by nearly three years of the reporting date.

The decision to seek funding on an agency-by-agency basis also created problems of imbalance since Congress seemed more inclined to support the power and flood control phases of the Corps of Engineers than the recreation and fish and wildlife investigations proposed by Interior. For example, as the question of reservoirs grew more pivotal, Interior made a valiant attempt to obtain separate appropriations to study the Corps-identified power sites. The proposal drew bipartisan fire from the Appropriations Subcommittee, despite its inclusion in the President's budget.

On the states' side, lack of manpower and appropriations fore-closed any meaningful input of state and local resources other than data already on file or available to participating state personnel. While private interests were represented on a number of subcommittees, only the utilities made any significant contribution to the studies under way.

Under these circumstances, it was inevitable that a further exten-sion of time and funds would be sought by some executive council members. Despite much justification for such a course of action, the recommendation was rejected by the chairman agency in favor of do-ing the best job possible with the resources at hand.

1953

There is some evidence that political factors contributed to this deci-sion. The Eisenhower administration assumed office early in 1953 with an outlook somewhat less liberal than that of its predecessor. A key individual in the new administration was former Gov. Sherman Adams of New Hampshire, never an ardent advocate of NENYIAC.

The New England utilities, in fact, gave serious consideration to stopping NENYIAC in its tracks at that time. By a curious paradox one of the leading private utilities, the New England Power Company, may have saved the day by urging completion of the study in order to provide the facts and figures on resources so badly needed in the region. This seeming inconsistency on the part of the utilities can be explained only within the context of the times.

It should be remembered that the New England utilities were al-most totally preoccupied with a backlog of construction accumulated during the war years and could give only cursory attention to the NENYIAC investigation. Moreover, the Charles T. Main report and the *Providence Journal Bulletin* series had reduced the pressure of pub-lic power proponents at least temporarily.

Even more significantly, industry studies were beginning to reveal that the technology of the future would be either larger steam gener-ating plants or nuclear fuels, with hydro power an increasingly doubtful component.

Finally, as NENYIAC progressed, interpersonal relationships be-tween utility participants, state representatives, and the civilian engi-neers of the Corps, grew increasingly cordial. There was mutual con-fidence as to both the objectives of NENYIAC and the sincerity of its participants.

Late in 1952 chairman Talley had made a tour through New England to appraise the utility climate firsthand. As he described later to the chief of engineers, Lt. Gen. Lewis A. Pick, the heads of the power

companies really "let down their hair"[36] for the first time, stating that
their greatest apprehensions lay in a possible domination of the study
by the Department of the Interior. Talley was quick to provide assurances to the contrary, adding that the Corps's only objectives were to
prepare a factual and unbiased report.

Mid-term in the NENYIAC study a number of central issues and problems had become evident. These were well summarized in a memorandum prepared for the incoming administration by Department of
the Interior representative Leland Olds. While it is not the intent of
this paper to dwell on personalities, Olds was a sufficiently central
figure in this early period to warrant at least a few explanatory
remarks.

A former chairman of FPC during the Roosevelt administration, and
then director of Interior's Division of Water and Power, Olds was regarded by some as the reincarnation of the devil but by most as a bril-
liant and impressive public figure. His bias toward public power
made him unusually controversial, as did his typically incisive line of
questioning at committee or public sessions. Yet of all the NENYIAC
participants, it was Olds as much as anyone who stretched the study
concepts beyond merely mechanical fact-gathering operations.

In an internal memorandum dated January 9, 1953,[37] he outlined
seven salient policy and procedural problems, observing at the outset:

the NENYIAC survey is significant as an experiment in integrated planning of
resource development to achieve a truly multiple-purpose approach. As such
it will be tested against such alternatives as the valley authority or the decen-
tralized Federal resources agency recommended by the Hoover Commission
on Organization of the Executive Branch of the Government.

The seven points included the achievement of unity in planning;
how conflicting uses of resources could be resolved; the manner in
which economic justifications and cost allocations would be handled;
the general matter of upstream flood control; the relative comprehensiveness of the various components of the investigations; the issue of
private versus public action in the ownership, development, and
utilization of resources; and the role of hydroelectric power under
federal auspices in competitively reducing private utility rates,
stimulating economic growth in underdeveloped areas, and providing reasonable control of New England river flows. His general appraisal of the NENYIAC effort was almost prophetic.

This would emphasize again that it (NENYIAC) represents a most important
laboratory experiment in the development of the Nation's resources policy. It
is particularly important because it is testing the possibility of accomplishing

desirable results through Federal-State cooperation in a region where, perhaps more than any other part of the country, there has been suspicion and resistance to any coordinated Federal program, particularly where hydro-electric power might be involved.

Inevitably, a number of thorny day-to-day issues had also arisen. The Department of Agriculture, for example, chose to undertake its economic and land-use studies within its traditional county lines, which had to be reworked to conform with river basin designations; agricultural and highway interests hotly debated the degree to which sedimentation was a problem within the study area; foresters and hydrologists argued vehemently over the extent of the influence of forest cover on flood and river conditions; the engineers held spirited discussions on the best method of determining flood frequencies; and the state representatives, particularly those from northern New England, reacted strongly to the growing implications of water quality control.

1954

But by far the liveliest battle was that waged by Interior over possible reservoir sites. By early 1954, Interior representative Abelson, who had succeeded Olds upon his retirement, reported that his agency was "waging and winning the fight"[38] for inclusion of alternate fish and wildlife plans in the individual basin reports. A year later, however, only one of the seven alternates had been formally accepted by the NENYIAC executive council. The exception occurred on the St. John River in Maine, where Interior's proposal would save the wild-life-rich Fish River lake system and the incomparable Allagash Wilderness country. Interior's lone success prompted a stiff response from the Corps, which did not appreciate having its engineering studies reviewed by another agency. The Corps also began quietly to solicit data on fish and wildlife and recreation from northern utility sources to counter the Interior estimates of economic value and public use.

By this time it was obvious that the report would never be written if each issue had to be met and resolved to everyone's satisfaction. A simple expedient thereupon evolved—to include both sides of an issue wherever possible in the final documentation. The policy was justified on the grounds that NENYIAC was to draw up an inventory of resources only and not an authorizing document. A prime opportunity to apply this principle involved the most fundamental issue of all, that of how hydroelectric power resources would be handled in the final report.

NENYIAC's power study and report group, under the chairmanship of FPC, had heretofore been unable to reconcile prevailing differences. The state members were generally opposed to any semblance of a public power report, whereas the FPC was desirous of including as many projects as possible in order to sustain its earlier estimates of a large potential of undeveloped power.

At the March meeting of the executive council, a special four-man committee was appointed to study the matter and return a recommendation at the next session. Included were representatives of FPC, the Corps, and the states of New Hampshire and Connecticut. Their solution, which was ultimately accepted by NENYIAC, was a creative one.

All sites with a power potential of more than 1,000 kilowatts and a storage capacity of more than 5,000 acre-feet would be tabulated by name, location, stream and drainage area, and listed in downstream order. For single-purpose projects, the costs would be determined on the basis of the organization most likely to construct them. In New England's case this meant private financing. For multipurpose projects, the costs charged to flood control and navigation improvements – admitted public responsibilities – would be computed on the basis of public financing. The power portions of such projects, however, would remain privately computed as before. The power values to be employed in the benefit-cost analysis would be predicated upon the cost of privately financed alternative steam generation, including also the cost of transmitting the power to those likely markets within economic transmission distance. Finally, in deference to changing conditions of power loads, construction, and fuel costs, the final inventory would include all projects with a benefit-cost ratio of 0.6 or higher.

With at least an uneasy truce now in effect, the Corps made another decision that further reduced the power issue within NENYIAC. By directive of the chief's office, a separate report was to be submitted by NED in compliance with Section 205 of the 1950 Flood Control Act. Although leaning heavily on the NENYIAC data, this would concern itself with the hydroelectric potential of the region's major river systems, as the original authorization had specified.

Coordinator Troland faced one other problem as NENYIAC began to draw to a close. President Truman's original directive echoed persistently in his ears. Like it or not, NENYIAC was to "prepare recommendations for specific action"[39] when the basic facts had been collected and analyzed. These instructions ran counter to NENYIAC's prevailing concept of an inventory only. Scars from the past could well reopen if specific project recommendations were to be included in the final report.

The solution devised was both legal and ingenious. NENYIAC's only recommendations as contained in the closing section of Part I of the General Report was that "the river basin and regional plan set forth in the report serve as a guide for the development, conservation, and utilization of the land, water and related resources of the New England–New York region"![40]

1954

By early 1954 it seemed prudent to begin considering what might be done with the final report. NENYIAC had already settled the procedure in its own mind, visualizing channels extending from the NENYIAC chairman to the chief of engineers, to the Secretary of the Army, and thence to the President and the Congress. Washington, however, had other ideas. The first inkling of them came in an advisory from the Office of the Chief of Engineers on April 19. Since the assignment of NENYIAC had been made by FIARBC originally, General Itschner said, it was more than probable that the submission should be made to the chairman of FIARBC. Either he or the federal agency heads jointly would then submit the report to the President.

NENYIAC's state members objected strongly. They most certainly did not want to run the risk of having their independent study findings watered down. Besides, the governors had made their appointments by request of the President, not of FIARBC, and the state members felt no obligation whatsoever to the latter agency. NENYIAC chairman Talley agreed. His response to the chief's office was concise and to the point. "I will state emphatically that I consider it to be against the best interests of the Corps of Engineers for this report to be submitted directly to FIARBC."[41]

Before the matter was resolved, FIARBC itself went out of business. By letter dated May 26, 1954 to the Secretary of the Interior, President Eisenhower gave his official blessing to a new executive branch Inter-Agency Committee on Water Resources (IACWR), replacing the informal FIARBC group.

NENYIAC, however, continued to press for some resolution of the question. The problem was shared by its counterpart for the Arkansas-White-Red river basins, which had been authorized at the same time as NENYIAC and was also chaired by the Corps. Accordingly, the representative of the Army brought the matter up for discussion at the September meeting of IACWR.

IACWR's decision was to reverse the previous FIARBC position. By letter dated November 24, 1954, chairman Stueck advised Talley that the final report should be submitted through Corps channels to the Secretary of the Army, who would then solicit official comments from the federal agencies and the governors of the states concerned before forwarding it to the President and the Congress. IACWR's good offices would be used, he added, only in the event of policy differences at the level of federal review. Thus vindicated, NENYIAC could devote full attention to its own endeavors. The June 30, 1955, deadline for submission seemed perilously close at hand.

By the end of 1954 sufficient material was on hand to begin the final round of public hearings.

The first was convened in Berlin, New Hampshire, on November 9.[42] Like many of those to come it was dominated by discussions of a particular project, the High Erroll Dam and Reservoir on the upper reaches of the Androscoggin, favored by the paper companies and area municipalities but questioned by fish and game interests because of probable loss of wildlife and fish habitat. A prominent witness was Dartmouth College, which owned some 27,000 acres of managed forest and wilderness in the general vicinity and also the last feasible impoundment site north of the proposed Erroll project.

The committee traveled to Augusta, Maine, the following day.[43] It was greeted warmly by Gov. Burton Cross at the public session. Again, the discussion related principally to potential water development projects identified in the NENYIAC report.

The chief engineer of the Oxford Paper Company described most of the projects as uneconomical to build and repeated his company's categorical opposition to what he termed "any socialization of the power industry." But a change in attitude had clearly transpired in Maine. This was well illustrated by the remarks by vice-president Harold F. Schnurle of the Central Maine Power Company, a previous NENYIAC opponent.

I might say that when this survey first started I, like many others, approached it with many misgivings and I might likewise say that under the able leadership of the chairman and the Council out of chaos has come order. I think that those who have actually taken the time to read the so-called 'Gold Books' could not help admit that they do constitute a real guide for the future in all phases of water resources in the area which has been studied.

Commissioner of inland fisheries and game Roland H. Cobb pointed out another benefit that had been derived from the NENYIAC studies – "this first opportunity for the States to make a detailed study of their various resources."

On December 14[44] NENYIAC convened the first of two hearings in Massachusetts, with a thin crowd of sixty-four persons attending. The tone of the testimony was again generally favorable.

Navigation and flood control interests testified to the need for prompt implementation of many of the projects identified by the study, and wildlife interests were out again in force, stressing the many discrepancies between the fish and wildlife task force recommendations and the final coordinated report for each river basin.

Potential conflicts between impoundments for power and flood control were cited by Hartford, Connecticut director of public works Charles W. Cooke, while Municipal Association representative Francis King spoke of a better competitive position for New England industry with the help of the lower-cost public power that might become available to the municipal utilities as a result of NENYIAC-identified projects.

An impressive delegation from New Haven, Connecticut, headed by Mayor Richard C. Lee and his renewal director, Edward J. Logue, greeted NENYIAC in that city on December 15.[45] The discussions focused on the role harbor improvements could play in the city's redevelopment program.

Equally significant, however, were the remarks of vice-president Robert P. Stacy of the Connecticut Light and Power Company. Admitting to initial skepticism, he was now of the opinion that NENYIAC had done "really a remarkable job." He reserved his particular kudos for the reports on hydro power, which he suggested would "clear the air of charges and countercharges and misinformation." Prophetically, he spoke of the rapid progress being made in the field of atomic energy and warned that such sources of power could make many of the undeveloped hydro sites less attractive economically than they were now.

Philip Barske, Northeast field representative of the Washington-based Wildlife Management Institute, himself a Connecticut resident, spoke of the general public apathy toward the committee's comprehensive work. He criticized the limited input from the states, commenting that if this reaction continued the only recourse would be regional resource planning under federal agency domination.

The following day found NENYIAC in Providence, Rhode Island,[46] for the fifth in the final series of public hearings. With the memory of Hurricane Carol still fresh, it was not unexpected that testimony would revolve around flood protection and navigation improvements within the Greater Providence area. Chairman Talley promised the group that the NENYIAC party would tour Narragansett Bay for a firsthand look at such problems following the close of the hearing. The Corps' inspection boat, the *Sea Echo*, was available for this pleasant duty.

A statement from Charles H. Callison, conservation director of the National Wildlife Federation, underlined again the dearth of public interest at previous public hearings. He cited the fish and wildlife development plan for the St. John River Basin in northern Maine, which embraced both power and the preservation of the Allagash Wilderness complex, as the only instance in which NENYIAC had clearly set forth alternative plans in its studies. This approach was suggested as a prototype for other river basin investigations.

1955

In Montpelier, Vermont, on January 12,[47] testimony was spare and to the point in the finest Green Mountain tradition. NENYIAC completed its business within the hour. A small delegation from the upper Winooski Valley expressed its interest in economic development and generally supported public power development projects within Vermont's "northeast kingdom."

Consulting geologist John R. Mills, appearing as a resident of Dorset, introduced a new thought: "I think it is extremely important that the Committee does not fold its tent like the Arabs and steal away, and that there be a central gathering agency for further information." His suggestion was to bear valuable fruit before the year came to a close.

The following day in Concord, New Hampshire,[48] Gov. Lane Dwinell was on hand to greet the committee. In his formal remarks the Governor stressed the nonpolitical nature of the investigations, the remarkable teamwork that had developed between state and federal personnel, and the emerging significance of the regional approach to resource problems in New England. The hearing brought forth tales of Vikings walking the beach at Great Boar's Head, a warm endorsement of NENYIAC from fish and game director Ralph Carpenter and a blunt statement from the Franconia Paper Corporation that any mandatory cleanup of paper mill wastes from the East Branch of the Pemigewassett River could jeopardize a $3 million local payroll.

The three public hearings in New York State[49] generated much heat but also some light. At Albany on February 8, defenders of the forest preserves were out in force to protest NENYIAC-identified power sites at Panther and Kettle mountains. State College (Albany) political scientist Robert Rienow lashed out at the invasion of states' rights, describing NENYIAC's consideration of such sites as "presumptuous," "a form of carpet-bagging," "an arrogation of Congress's own powers," and a "man-handling of the delicate line between Federal and State relations." A strong case was made for the further deepening and widening of the navigation channel from Albany to the mouth of the Hudson River. And national fish and wildlife representatives termed the inventory phases of the report "a milestone and a project to be commended," but panned the coordinated plan as leaving "a great deal to be desired."

The great and the small pleaded for more time to examine the studies in depth, but time was clearly at a premium if NENYIAC was to submit its report to Washington by the projected March 15 deadline.

Panther Mountain padded back into the spotlight in Syracuse, New York, the next day, and Syracuse University geographer Eleanor E. Hanlon spoke eloquently of the need for an ethical sense in people.

It was clean streams (the Niagara River in particular) and the eroding Lake Ontario shoreline at the hearing in Buffalo as the third and final New York State hearing came to a close on February 10. A topic of some speculation was the likely level of Lake Ontario once the St. Lawrence Seaway and Power Development Project construction was completed. John L. Beyer, former mayor of the City of Tonawanda, suggested an incentive program to enable industry to charge off the cost of waste treatment measures against the federal income tax at a rate of 20 percent a year.

By February 1955, Colonel Troland had done his work so well that consideration could be given to closing down the NENYIAC central office. Official copies of the final report were to be mailed out on March 15 and all records and files were then to be transferred to NED. As of March 31 the Boston office would be closed.

NENYIAC's final public presentation took place in Boston on March 10.[50] Joseph F. Sharp of the American Pulp and Paper Association expressed his organization's concern over the unrealistic and impractical nature of the report's industrial waste treatment sections. The association's position was that stream improvement belonged properly to the states and interstate agencies, not the federal government.

From several New Hampshire spokesmen came a provincial echo from the past. A New Hampshire House of Representatives resolution was introduced into the record urging that the initiative in the development of the natural resources of the state should remain within the state. Concern over what might happen after NENYIAC expired was expressed by both Harold D. Resseguie of the Public Service Company of New Hampshire, and state planning director Sulo J.

Tani. A live NENYIAC appeared clearly preferable to an unknown future.

The New England Council, the region's most prestigious industry organization, put in its first official appearance. "New England has always been a do-it-yourself area," said spokesman Melvin D. Peach. "We can all do things better when we have the facts to work with. This inventory of resources should be extremely helpful to New Englanders in their planning of the utilization of these resources."

The clerk of the Connecticut River Valley Flood Control Commission, former Vermont attorney general Alban J. Parker, spoke apocryphally of John the Baptist who lost his head to a harlot. Not so the commission, he said, because several of the NENYIAC flood control sites in Vermont and New Hampshire were still to be accepted by his Interstate Compact Commission. Nevertheless he urged prompt action on those reservoirs where mutual agreement existed, for no construction had been started on flood control dams in the Connecticut River Valley for many years. "The need is still existent and is perhaps more urgent than any of us today realize." His point would be underscored dramatically in the Hurricane Diane floods of August 1955.

In his closing remarks at the public hearing, chairman Talley spoke of many things—the approximately 145 full-time staff participants who made the report possible, the $5 million cost of the survey, the 200 pounds of final reports, and the nearly 1,000 people who had contributed in some way to the undertaking.

It is a study in which the private individual, private industry, private organizations in conservation, and in business and economics, have played perhaps a greater part than in the conduct of any similar activity, and foremost in its uniqueness is the fact that it is a survey in which the States have had and have exercised a voice of co-equality with the Federal representatives on the Council. . . .

We have not recommended in this survey that a specific thing be done by any specific agency. Rather we have made an inventory of the resources to the best of our ability . . . the area is too big, the resources are too big for a single agency such as this to have made any presumptions of that kind. . . .

I believe the most unique thing that has come out of this survey has been the demonstrated ability of the private individuals of the States and the Federal Government to work together on a team, on a plane of co-equality. . . .

We do hope, however, as someone has said, that this is not the end of the survey. We do not believe that it will be. Its future quite properly rests in the hands of the people of this great region, just as the future of these resources rests in the hands of the people.

Dudley Harmon, in his *Providence Journal* commentary, wrote, "It is impossible, in a few paragraphs, to convey an adequate idea of the contents and values of this report. It is unprecedented in its magnitude and scope."[51] He reported that Gov. Dennis Roberts of Rhode Island, chairman of NEGC, would place before the next meeting of the governors a proposal for a continuing body to stimulate and record action on the innumerable findings of the report.

With the end appreciably in sight, an unexpected reversal suddenly faced NENYIAC. IACWR had changed its mind on how the final report was to be handled. The Bureau of the Budget had stepped in with a new procedural suggestion whereby the heads of the participating agencies would jointly transmit the report to the President and, upon receipt of his comments, would transmit the final document to the Speaker of the House and the President of the Senate.

A minor revolution took place at the NENYIAC executive council meeting on March 10. Vermont representative Shutler echoed state sentiments when he spoke of the executive council reorganization as having given NENYIAC "an entirely new character" and "an example of cooperation in being and in action."[52] Now that the need for cooperation has ended, he said, it appears that the federal agencies desire to take over once more.

In responding to IACWR chairman Stueck, NENYIAC chairman Talley urged a change of heart so that "the fine relationships established among the Federal and State representatives in the region may be preserved and that the tremendous progress made toward Federal-State cooperation in resources matters be not endangered."[53]

At its May 10 meeting in Washington IACWR finally capitulated, returning to the previously agreed upon method of submission. Since IACWR was visualizing a series of field interagency committees across the country, one of which would encompass the New England–New York region, it seemed imprudent to offend a group of potential sponsors at this time.

Secretary of the Army Robert T. Stevens had written to each of the governors on March 30, 1955 and to the heads of the federal agencies on April 12 requesting formal comments on the NENYIAC report by June 15. Always mindful of its congressional relations, the Corps also advised members of the New England delegation that the study had been completed.

The individual responses made much of the comprehensiveness of the report and the fact that it was an inventory rather than an authorizing document. Such terms as *exhaustive, voluminous,* and *thorough* were used to characterize its preparation and contents. NENYIAC was described as starting badly but ending well. It was given particularly high marks for the democratic way in which it operated. The agency comments underscored several obvious weaknesses in the report, e.g., lack of consideration of power redevelopment possibilities, adverse effects on fish and wildlife, and the implication that no water shortages would occur in the next fifty years. Interior felt that alternative proposals should have been included (especially its own) or at least a minority section provided for dissenters in the final report. Three of the governors spoke of the need for a continuing organization of some description to carry on NENYIAC's work and to see that the sizeable investment in resource studies was not wasted.

As it later developed, much of NENYIAC's material would shortly be out of date, for as the final comments were being incorporated in a set of so-called black books, hurricanes Connie and Diane descended with unusual devastation on New England, bringing floods of record to most of its south-central streams. As the chief of engineers described the event to IACWR, the August storm pattern dropped as much as 18 inches of rain in a 36-hour period within portions of the region. He estimated property losses on the order of $1.5 billion— some 3 times greater than the November 1927, March 1936, and September 1938 floods of record put together. Some 200 lives were lost before the flood waters receded.

On October 10 the governors were asked again for comments. Flood control was understandably on their minds. By extraordinary coincidence, a special committee of NEGC under Massachusetts commissioner of public works John A. Volpe had been working for the past three months to obtain a higher share of the annual federal river and harbor and flood control appropriations, feeling that New England's 1 percent of the nearly $5 billion available in most years was hardly consistent with the extent of its tax contributions. By the advent of the hurricane rains, NED engineer Robert J. Fleming, Jr., had already furnished the committee a complete breakdown of projects needed in order of priority.

At the governors' meeting in Boston on September 23, project requests totalling more than $47 million were approved and authorized to be submitted to the President by chairman Roberts of Rhode Island. Governor Roberts also included these estimates in his supplemental comments on the NENYIAC report.

New England's fortunes in seeking flood relief funds are worth pursuing for a moment, tangential though they are to the thread of the NENYIAC story, for they do illustrate a problem the region was to experience later in using its full weight to advantage in Congress. Remedial efforts proceeded on a number of different fronts. Prof. Seymour E. Harris, chairman of the New England Governors Textile Committee, put in a strong pitch for New England flood relief and flood insurance before the hearings of the Subcommittee on Water

Resources (House Government Operations Committee) in Springfield, Massachusetts on October 25. A special committee of the Massachusetts congressional delegation produced a flood report entitled
"Disaster Strikes," which was circulated extensively within Congress.
Sen. Prescott Bush of Connecticut attempted to persuade budget director Rowland Hughes to part with a major share of the President's
$1 million emergency fund for needed reservoir construction projects
in New England. The stumbling block appeared to be not Hughes,
but House Appropriations Committee chairman Clarence Cannon,
who remained unmoved by the urgency now that the floods were
over.

1956

Failing in these overtures, the New England group joined forces with
New York and Pennsylvania in seeking special deficiency legislation
in the 84th Congress. On February 3 the National Rivers and Harbors
Congress reported that the deficiency bill for fiscal 1956 would be
some $3 million below President Eisenhower's request. The only proj-
ects cut were those sought by the New England–Appalachian coalition! Gov. Abraham Ribicoff of Connecticut was dispatched to Washington to speak for his fellow governors when the House floor action
took place. As the New England Council's executive vice-president,
Walter Raleigh, described later from a seat in the gallery, Ribicoff's access to the floor, plus Louisiana Congressman Overton Brooks's willingness to sponsor a restorative amendment, finally carried the day
for the New England cause.

If the governors had been only lukewarm about water resources in the
past, the combination of NENYIAC and hurricane floods had now
made natural resources a significant issue for them. At least two briefing sessions had been held in the course of NENYIAC, and they were
reasonably current on its status. At the March 15 meeting of the governors in Boston, the forthcoming report was discussed at some
length, and it was felt that some appropriate group should be asked
to prepare a unified set of comments from a regional viewpoint. In
the absence of any more likely prospect, the matter was turned over
to the Governors' Permanent Regional Committee on Industrial and
Development Problems.
   This was a committee established in 1953 in response to the continuing problems of the textile industry. It consisted of two members
from each state chosen from their respective industrial and development organizations. Its current chairman was John Pillsbury of New

Hampshire, vice-president of the Public Service Company of New Hampshire and chairman of the State Development Commission. Inappropriate though the choice might appear, the event was most fortuitous, for Walter White of NENYIAC and John Pillsbury of the Governors' Permanent Regional Committee were close personal and political associates.

THE QUESTION OF A SUCCESSOR

NENYIAC, in the meantime, had given careful thought to the matter of a successor organization. New Hampshire representative Walter White, with the assistance of Connecticut's Richard Martin and Vermont's Philip Shutler, presented a draft charter to the final executive council meeting on April 14, 1955. It provided that the chairmanship be rotated among state and federal members, and a central office and staff be maintained at joint state and federal expense. Following much discussion, the executive council was able to agree only on the need for a continuing body of some sort, a geographical jurisdiction consisting of New England and New York, and a council of state and federal representatives on a coequal basis receiving what the council referred to as "appropriate staff assistance."[54] With this degree of endorsement, the state-drafted charter was forwarded to IACWR on May 6, 1955.

IACWR's response on June 17, 1955 was friendly but firm. Chairman Roderick gave general assurances of participation and interest but stated that the staffing and financial commitments on the federal side could not be entertained at this time. He returned a further redrafted version of the charter for New England's consideration, which yielded only on the point of coequal representation.

As a federal observer later reported, NENYIAC's own meeting on July 20 prior to the joint session was spirited. The New Hampshire and Vermont representatives led the way in discussing procedures for establishing a successor organization. Other state and federal representatives had few comments of any import, the federal members, in particular, feeling that the states should "carry the ball." Though there was general agreement with the IACWR-prepared charter, Shutler and White were not prepared to abandon the office and staffing provisions they felt were essential. White suggested a special committee appointed by the President and the governors to negotiate a satisfactory instrument between the federal government and the states. To some of the federal representatives present, this insistence was difficult to understand in the light of the very substantial area of agreement with most sections of the IACWR document.

The joint meeting with the Governors' Permanent Regional Committee on July 21, 1955 produced a meeting of minds. It was determined that the governors would be approached at their August meeting with White's recommendation for an appointed presidential-gubernatorial committee. Temporary chairman Shutler confidently predicted that a permanent Northeastern Resources Council would be in business by November.

Unfortunately, it was not until their November 17, 1955 meeting in Boston that the governors found the time even to consider NENYIAC's suggested approach. Walter White of NENYIAC told them that the region was losing personnel and information rapidly because there had been no continuation or implementation of the NENYIAC studies. Unless a permanent Northeastern Resources Council was established, he said, the states would undoubtedly return to the unfortunate business of competing against each other for federal public works projects.

Governor Roberts of Rhode Island, the conference chairman, asked whether the federal government would support its share of the proposed council. White admitted that there was a question on this point at present and suggested that the council be set up on an informal basis by the governors for a year or two and then, if shown to be successful, formalized by interstate compact.

White also advised the governors of the proposed IACWR charter, stating that it did not call for coequal representation and was highly controversial among the federal-state representatives. He was challenged on this overstatement by Governor Roberts, who questioned whether there was really sufficient disagreement among NENYIAC members to warrant the special committee sought by White.

Not entirely convinced of the merits of the proposal, the conference voted to have its chairman explore the attitudes of the federal government and the State of New York before taking action on the proposal.

Governor Roberts carried out his instructions of November 29, 1955, requesting clarification from IACWR chairman Stueck on three specific items: the matter of coequal representation, procedures acceptable to the federal government for administration and finance, and possible participation by the State of New York. Stueck's response was noncommittal in the light of a new development on the Washington scene. The report of a Presidential Advisory Committee on Water Resources Policy was expected momentarily. When it was submitted to Congress on January 17, 1956, copies were sent to the New England governors, and Governor Roberts was advised that IACWR would consider the New England situation in the light of these policy recommendations.

Much of what was recommended in the report made sense to New England. The central theme was that of cooperative federal, state, and local water resources planning on a continuing basis with regional or river basin water resources committees as the instruments by which it would be achieved. Sensitive New England antennae spotted two particularly desirable recommendations: coequal representation among state and federal designees, and an endorsement of the interstate compact approach to water problems. A third recommendation was not so desirable, but probably inevitable: the chairman of each committee would be a presidential selection. There was substantial amelioration, however, in that the chairmanship would be independent of any federal agency and nonvoting in character.

With this information in hand, Governor Roberts met with his fellow chief executives in Boston on March 26, 1956. By motion of Governor Johnson of Vermont, seconded by Governor Herter of Massachusetts, the conference agreed to seek the establishment of a resources council of New England patterned after the advisory committee report. Governor Roberts was to contact IACWR in this regard and to approach New York again for possible participation in a coequal federal entity for the Northeast.

On June 30, 1956, IACWR chairman Roderick informed Governor Roberts that he and his associates had formally adopted a charter for a Northeastern Resources Committee. Federal designees would be named shortly. Roderick tactfully deferred on the specific steps to be taken by the governors, other than to suggest several ways by which confirmatory action could be taken by the states.

In the final chapter of the charter, seven federal agencies and seven states (including New York) were listed as eligible to participate in the work of the committee. Coequality was specified in a paragraph early in the document. The chairman was to come from state and federal members as open elections would determine. On the matter of budget and staff, however, the NENYIAC proposal was not accepted. Other than the secretarial services provided by the chairman agency, and any additional staff assigned to it by member agencies, NRC was to have no direct personnel or budgetary provisions. IACWR, however, in its letter to Governor Roberts did dangle the prospect of some more permanent committee—"if and when such action is possible"![55]

By the time of the governors' meeting in Newport, Rhode Island on July 16, 1956, Governor Roberts could tell his fellow chief executives that New York had expressed interest in joining NRC. By September, all seven federal representatives had been appointed.

On November 15, 1956, the Corps, Interior, and IACWR representatives journeyed to Boston to meet with the New England governors.

Provisions of the charter were discussed, as well as procedures still required on the governors' part.

At the IACWR meeting on February 12, 1957, tentative plans could be reported for the organizational meeting of NRC. All the New England states, with the possible exception of Massachusetts, were expected to be present, and the latter was merely delinquent, not uninterested.

NRC's long and difficult gestation period appeared to be almost over.

CHAPTER THREE

The Northeastern Resources Committee

1957

The first formal meeting of the Northeastern Resources Committee took place on February 25, 1957 in Boston and was attended by Governors Johnson of Vermont, Muskie of Maine, Furcolo of Massachusetts, and Roberts of Rhode Island, and by Assistant Secretary of Defense (Army) for Civil-Military Affairs George H. Roderick, who conveyed the official blessings of the Interagency Committee on Water Resources.

Roderick spoke with surprising candor of some of the problems troubling NRC's state members, characterizing the valley authority device and the reorganization of states into political counterparts of river basins as steps that were now clearly out of touch with the times. He advised active participation by local and state interests to prevent domination of NRC by federal agencies, adding obliquely, "Willingness to share the costs is the best indication of the merit of a project."[56] Roderick's advice was soon to prove prophetic.

When the governors had departed, NRC set to work. Walter White of New Hampshire was elected the first chairman, and to Philip Shutler of Vermont, the other leading architect of NRC, fell the task of heading a special committee to prepare appropriate by-laws and program recommendations.

At the April 11 meeting in Boston, the Shutler committee presented a draft set of by-laws. These were debated spiritedly in traditional NENYIAC style, and many changes were made.

Fresh from a gaggle of NENYIAC task forces and subcommittees, NRC gave particular attention to its own committee structure. Only two were established initially: publicity and operations. With Chairman White an ex officio member, the states were thus assured of a majority position on NRC's first executive committee.

Included among the approved by-laws were a number of benefits NRC was expected to produce for the region: improved coordination; resolution of conflicts between agencies; adjustments of conflicts in interests; and adjustments of state and federal agency policies and programs to meet regional needs. Overall, NRC was to develop a program for the progressive development and use of the region's resources that, in the committee's words, "could and would be supported by a great majority of the people and businesses and by the states of the region."[57] As initial steps toward this objective, the Shutler committee recommended three program items: 1) a general plan for the coastal area of New England; 2) a plan for reforestation of New England; and 3) a plan for region-wide flood protection.

Since NRC had no prospect of either staff or funds, an approach was made to private foundations. At the June 13 meeting, it was reported that neither the Rockefeller nor the Conservation Foundation in New York was overly optimistic, but each invited a formal request for assistance. NRC later submitted applications, but these were rejected, the opinion of the foundations being that NRC should be supported by public appropriations.

NRC also heard from its Publicity Committee. At the suggestion of chairman D. R. Gascoyne of Interior, a series of public meetings in each state was agreed upon with two objectives: 1) to acquaint the public with the NENYIAC findings; and 2) to bring NRC's program and objectives to public attention.

Also at this meeting, chairman White expressed his growing concern over the lack of an official document that properly expounded NRC's credentials. The charter as issued by the Inter-Agency Committee on Water Resources (IACWR) made no mention of the actions of the New England governors in authorizing NRC's establishment. White felt that the duality of NRC's existence was particularly significant and should be made very clear at the various state meetings. NRC agreed, and the Corps of Engineers offered its legal drafting services for the preparation of a suitable document.

In the meantime, the draft memorandum of understanding, prepared for the signatures of the New England governors and the members of IACWR, had encountered various shoals. Faced with an unprecedented situation, IACWR had responded bureaucratically that the memorandum seemed to constitute an interstate compact requiring the consent of Congress.

At the New England Governors Conference in Boston on November 21, when doubts were expressed about a governor's authority to sign, the governor of Maine offered to have his state legal officer draft an acceptable letter of agreement for circulation among all the governors.

1958

Designed to precede the formal language of the NRC charter as approved by IACWR, the document was presented to the New England Governors Conference (NEGC) in Stowe, Vermont on January 20.

Some governors felt strongly that the document should in no sense commit the states in advance to any position advocated by NRC, and there was also unanimous agreement that any reference to funds should be stricken, since official legislative sanction would be required for budgetary support. Thus emasculated, and still subject to further rephrasing, the memorandum was finally approved in principle by the governors.

It required a rump session of NEGC, held May 17, 1958, in Miami in conjunction with the National Governors Conference, to approve the agreement finally. The records indicate that Governors Johnson of Vermont, Muskie of Maine, Dwinell of New Hampshire, and Ribicoff of Connecticut attended this meeting. Governors Muskie and Dwinell actually worked out the final language in long hand during the course of the meeting to be certain that the states could act independently of NRC. At long last NRC had its official credentials.

Despite this subdued and calculated support from the New England governors, NRC's spirits were far from dampened, for there was genuine enthusiasm from others for its projected schedule of state meetings. Resources economist Robert W. Eisenmenger of the Federal Reserve Bank of Boston consented to tackle the difficult job of compressing the NENYIAC report into individual state summaries. He would serve as lead-off speaker at each public session to present the facts. Francis E. Robinson of the New England Council's interstate relations staff agreed to handle advance publicity. The Council's stature and visibility within the region was bound to attract interest from the private sector. Should NRC ever acquire a staff, it also had offers of office space free of charge from the New England Division of the Corps of Engineers and the University of New Hampshire. Moreover, an important milestone had just been reached. Almost three years after submission, two sections of the NENYIAC report had just been issued by the Government Printing Office as Senate Document 14 of the 85th Congress, 1st Session. To veteran NENYIAC participants, this was heady news indeed!

Thus buoyed in spirits, NRC headed into its first state public meeting. Nearly three hundred citizens crowded into the Highway Hotel in Concord, New Hampshire on January 30[58] to hear Governor Dwinell extend his personal greetings. Chairman White described NRC's hopes and aspirations; and the principal NENYIAC findings for New Hampshire were summarized.

NRC had correctly gauged the temper of its audience by assigning the largest portion of available time to concurrent panel sessions where citizens were free to speak their minds. Each session was chaired by a well-known and respected individual drawn from the nongovernmental sector—in the case of New Hampshire a farmer, a landowner, a prominent legislator, and a leading businessman. To Walter White, determined champion of the NRC cause, the turnout was particularly gratifying. As his associate from Connecticut, William S. Wise, stated publicly, "Walter had a dream . . . and it was realized today!"[59]

With some variations, this pattern was followed at each of the subsequent state public meetings: Providence, Rhode Island (March 21); Hartford, Connecticut (April 14); Amherst, Massachusetts (May 26); Waterville, Maine (June 11); and Montpelier, Vermont (June 29). Time and space permit highlights from only a few of these sessions.

Associate Director Joseph L. Fisher of Resources for the Future, Inc., Washington, D.C., told his Providence[60] audience that it must give thoughtful consideration to the problems of resource development throughout New England. As a former staff associate of the National Planning Association's Committee of New England study during the early 1950s, Fisher was eminently qualified for this assignment.

Walter N. Phillips, executive director of the Delaware River Basin Advisory Committee, told the public session in Hartford[61] of the similar frustrations of his group in trying to promote regional action. He traced the historical growth of federal interest in water resources, commenting that self-determination becomes even more difficult in the face of the twenty-five separate federal agencies already charged with water resources functions by Congress.

Massachusetts participants at the Amherst[62] session were privileged to hear from two keynote speakers: internationally respected sanitary engineer Abel Wolman of Johns Hopkins University (a friend of the Commonwealth's NRC representative Clarence I. Sterling, Jr.) and dynamic Massachusetts Commissioner of Natural Resources Francis W. Sargent, proponent of a then revolutionary $100 million outdoor recreation program for public land acquisition and park development.

NRC was correct in assessing the public meetings a success. More than one thousand individuals attended—a significant feat in itself— and the scope of the discussions revealed great interest in the subject at hand. Yet, as history would disclose, the public somehow failed to grasp the significance of NRC itself, and what it interpreted to be support may well have been only a growing general interest in the natural resource field as a whole.

Absorbed though it was in its state public meetings, NRC could not neglect for long its own internal operations. In fact, having to arrange the state sessions with only volunteer manpower had emphasized again just how limited NRC was without central funds and staff.

The situation came to a head at the executive session in Hartford, Connecticut in April. Chairman White's feeling was that an interstate compact, sanctioned by the state legislatures and by Congress, was the only practical means by which funds from several sources could be mingled. The suggestion triggered prolonged discussion.

Ardent states-rightists warned of possible changes in national policies and procedures that could affect New England drastically if it was to become dependent upon federal funds. It would be far better to keep the compact a New England venture only, they contended. Yet other state representatives stated flatly that they could not assume their turn at the chairmanship without some sort of staff and financial support. This was estimated to cost $20,000 to $30,000 annually if the job was to be done correctly.

Federal agency representatives expressed little support for an interstate compact, arguing instead for the chartered organization now in being. It was pointed out that funds could not be provided directly to NRC by any agency, although support in kind was a possibility for some. White sought consensus by asking each member to state his preferences between a charter or a compact. The state representatives indicated general concurrence with the compact approach. Somewhat reluctantly, the federal members agreed that a compact would be a satisfactory solution but hedged their final support on the position to be taken by their peers on IACWR.

A specific course of action had to be determined. Upon motion of Gernes of Labor, seconded by Abelson of Interior, a special committee consisting of Sibley (Army), Watson (Agriculture) and Wise (Connecticut) was appointed to look into the matter of an interstate compact. There were immediate objections as to the preponderance of federal members, and the motion was withdrawn. A new Committee on Compacts was then appointed, consisting of Wise as Chairman, Tani (Maine), Thieme (Vermont), Martin (HEW), and Watson.

Thus, for Wise, career engineer and warmly respected director of the Connecticut Water Resources Commission, an assignment began that was to extend for almost a decade. He was to pursue this quest for a compact with utter dedication and conviction, but without appreciable success. It is ironic to note that this appointment to the committee chairmanship actually came in absentia!

Consensus within NRC on the compact approach proved short-lived, as the executive session in Amherst in May revealed. Concern was expressed about feasibility of the proposed compact; the absence of coequality; the lack of provision for future membership of other states; and the problems of financing. The Maine representative went so far as to propose an entirely new arrangement to be endorsed by the New England Governors and financed exclusively through their executive offices.

Committee chairman Wise told NRC that the several existing precedents—notably the Delaware River Basin Advisory Committee and the Interstate Commission on the Potomac River—fell considerably short of what New England appeared to need. As basic criteria, he noted, a suitable instrument should include a permanent secretariat, provide a source of funds for continued operations and, above all, be acceptable to all the states.

Following considerable discussion, it became obvious that the existing NRC charter was the only conceivable instrument on which federal and state representatives might possibly agree. Accordingly, the Committee on Compacts, with William S. Wise as chairman, was instructed to draft a compact along these lines, subject to possible modification at subsequent meetings. The federal representatives, however, made it plain that if an interstate compact was to be the course followed, the states must take the lead.

At this point, New England's considerable experience with interstate compacts, and its fortunate choice of a compact committee chairman, came into play. Wise was already serving on three compact commissions and was also an active participant in the Interstate Conference on Water Problems, an annual project of the Council of State Governments. He was instrumental in finding two skilled consultants, Dr. Mitchell Wendell of the council and Prof. Frederick Zimmerman of Hunter College, each of whom expressed great interest in helping New England draft an appropriate compact. Shortage of funds was overcome by an offer from the Corps of Engineers to make its legal staff in Boston available for drafting purposes, and Wendell and Zimmerman became available through the good offices of the New York Joint Legislative Committee on Interstate Cooperation.

What was especially intriguing to Wendell and Zimmerman was the active interest of the Corps in the compact drafting. Anxious to avoid the precedent of the Missouri Basin Compact, which had been disavowed by the Corps and subsequently buried before it even reached the light of day, the consultants made every effort to retain the language suggested by the NED staff attorney.

At the NRC session in Boston in September, a draft of the proposed compact was available for general discussion. Dr. Wendell told NRC frankly that it could expect certain difficulties at the federal level, since all previous compacts had been agreements between the states without full federal participation. He singled out the Department of Justice as a major likely objector.

Federal NRC members advised close consultation with IACWR during the drafting phases. IACWR representative George R. Phillips of Agriculture warned against any expectations of prompt Washington action. The state representatives countered by observing that all of the state legislatures would be in session in 1959—hence an ideal opportunity to obtain compact ratification. With January just around the corner, time was clearly at a premium.

On October 14 a special delegation of NRC met with IACWR at HEW in Washington, with special observers present from the Bureau of the Budget and the President's water resources development staff.

Following William Wise's presentation on behalf of the NRC delegation, various IACWR members raised the anticipated constitutional and policy questions, noting that what NRC was proposing was essentially a joint federal-state compact. Alternatives were suggested: 1) the creation of a federal regional water resources planning committee along the lines of the recommendations of the Presidential Advisory Committee on Water Resources Policy; or 2) the establishment of two entities, cooperative in nature but legally separate, one a federal commission authorized by Congress, the other an interstate board established by compact among the states.

The NRC state members told IACWR flatly that neither approach would be acceptable. IACWR Chairman Aandahl warned NRC that it would be unrealistic to expect compact legislation of this unusual nature to go through Congress the very first time. White responded by observing the tight schedule many of the states were facing because of the biennial nature of their legislatures. Recognizing the unlikelihood of any agreement during the IACWR session, Aandahl offered to appoint a special subcommittee to work with NRC on the compact issue.

Following the meeting the NRC delegation was unexpectedly invited to the White House to explore the matter further with Special Presidential Assistant James S. Bragdon. He considered the purposes of the compact highly desirable but strongly advised NRC to make its peace with the Department of Justice before moving to the stage of legislative enactment.

At the NRC session in Boston in October 1958, compact committee chairman Wise reported in detail the results of his Washington expedition, and there was uniform agreement on the need for an immediate Department of Justice review. The alternative of a regional water resources planning committee, as recommended in the Presidential

Advisory Committee "blue book," was also discussed, but the state members objected strongly to the appointment of a federal chairman as giving one man tremendous power over a region.

1959

Despite a sympathetic attitude on the part of the Washington agency representatives and concessions in the compact redrafted by the IACWR subcommittee, NRC again rejected the draft as not going far enough. It instructed its Committee on Compacts to prepare a final version suitable for submission to the various legislatures and eventually to Congress.

Compact committee chairman Wise forwarded the end product to Washington on January 30, 1959. At its February meeting, IACWR noted that it differed in basic principle from its own drafts and concluded that no further comments should be made until the time for consideration by Congress.

The persistence of NRC's state members – amounting to stubbornness and even arrogance at times – is clearly apparent throughout the official proceedings. NRC's unwillingness to meet the tenor of the times may well have cost it a half decade of significant activity, as the subsequent record will indicate. However, history was later to bear out many of its convictions. The federal posture of unconstitutionality was to be decisively broken with the establishment of the Delaware River Basin Commission in 1961, and the Water Resources Planning Act of 1965 was to preserve in legislative policy the wisdom of a co-equal partnership for water planning between the federal government and the states. Viewed in historical perspective, the seemingly provincial attitudes of the New England group may have been more responsible than many people at the time were willing to admit.

In the meantime, NRC's various state sessions had begun to earn important dividends, and in December NRC solicited and received assurances of active backing from the New England Council for its interstate compact proposal within the respective states and the Congress.

In order to expedite introduction and passage of the state enabling acts, it was decided that a formal presentation should be made to the next New England Governors Conference. Accordingly, on March 2 Chairman White appeared before the governors in Hartford, Connecticut to seek their support for the compact. White stated that the present NRC arrangement was weak because there was no provision for financing the organization. He outlined a compact that called for an annual budget of $100,000, split fifty-fifty between the federal government and the states. It was understood that the states collectively would contribute a maximum of $50,000 and that no state would be liable for more than its share, based on population and land area, regardless of how many joined. Provision was also made for membership of New York, in which case the financial apportionments would be adjusted. When three state legislatures had adopted the compact, NRC intended to file legislation for the needed congressional consent.

The combination of preliminary contacts by state officials, support by the New England Council, and persuasive presentation by chairman White carried the day with the governors, and the conference unanimously voted to support the proposed compact. All signals were clearly go!

By March 5, when the NRC met in Boston, compact legislation had been entered into the legislatures of Connecticut, New Hampshire, and Maine, and similar action was expected momentarily in Massachusetts and Rhode Island. NRC had the support of the Council of State Governments and its Joint Federal-State Action Committee, the National Association of Attorneys General, and the New England Council, which had contacted the entire New England congressional delegation. Before the year was well along, however, enthusiasm would have to be tempered by reality.

In Vermont and Connecticut, for reasons of economy, there would be no appropriation, and Massachusetts reported difficulty in even getting the governor to submit the required special message. In Maine timberland and utility interests opposed the compact for fear of federal domination, and opposition from its Forestry Committee for the same reason forced the New England Council to modify its support.

By October four states had enacted the compact legislation, but only two (New Hampshire and Rhode Island) had included appropriations. The Maine and Vermont prospects were dead until the next biennium.

In comparing notes, the state NRC members could identify at least one promising new influence in the region, the League of Women Voters, which had translated its national study item on water resources into effective interest and support for the interstate compact.

By 1959, NRC could detect some encouraging signs of progress on its interstate compact project. However, this preoccupation with a single issue was not without disadvantage. Matters seemed to be bogging down at home, and several innovations and rearrangements were attempted, among them those suggested by the new chairman, Alvin C. Watson of Agriculture. On his recommendation, an extensive series of committees was launched, and a deliberate effort was made to involve specialists from industry and agencies outside the regular NRC family in these assignments. By virtue of his representation of Agriculture on a number of other river basin committees, Watson was also extremely helpful in keeping NRC posted on events elsewhere. The discussions over a Delaware Basin compact, for example, were of intense interest and significance to NRC.

Efforts were also renewed to persuade New York to become a regular NRC member. The motives were partly cordial and partly calculated, for New York's large congressional delegation could become a powerful ally on behalf of the federal consent legislation. Although Governor Rockefeller responded to the invitation in positive terms, no New Yorker ever appeared at NRC meetings.

NRC members also began to wonder out loud about the validity of a major plank in their compact platform—the need to implement the NENYIAC findings. After all, much of the data was now almost a decade old. Discussions began on the desirability of focusing the attention of the various subcommittees on an updating of the NENYIAC report.

Meanwhile, the national water policy pot was boiling again. The federal members of NRC reported the appearance of the Administration's new legislative proposal, filed by Congressman Wayne Aspinall of Colorado, chairman of the House Interior and Insular Affairs Committee, on behalf of the Administration. Titled the Water Resources Planning Act of 1959, it contained several provisions that NRC members believed could seriously affect and even forestall passage of the compact in the various state legislatures, and an overall philosophy of federal dominance drastically counter to that expressed in the New England compact proposal.

The Senate Select Committee on National Water Resources (the Kerr Committee) was also beginning its deliberations, and NRC chairman Watson was invited to present testimony at the Boston hearing on December 2. A formal invitation to testify was also extended to NEGC, and each of the state water resources officials was encouraged to appear at the hearing. From this series of concurrent and fortuitous circumstances, New England gained a timely appearance before a respected and powerful audience. NRC participants used the occasion well to press their convictions regarding coequality, and to underscore the need for a federal-state compact agency in the Northeast. There is reason to believe that Sen. Robert S. Kerr of Oklahoma, one of the most influential members of the Senate at that time, listened attentively to the New England arguments as he presided, for the committee's later recommendations concerning water resources planning differed drastically from those presented earlier to the Congress.

1960

At the meeting in Boston in February, compact committee chairman Wise was delighted to report the introduction of a federal consent bill into both branches of the 86th Congress. The companion bills were S. 2842 by Sen. Prescott Bush of Connecticut, and H.R. 9999 by House minority leader John W. McCormack of Massachusetts. Rep. Chester Merrow of New Hampshire had also filed a compact bill in the House (H.R. 10022). All three had been referred to the respective Public Works Committees for deliberation.

Phase Two of the compact quest was now officially launched.

Within a week of the firing of the first legislative shot, the federal IACWR members met in Washington to examine the provisions of the interstate compact proposal. After reviewing the chain of events of the past year, the consensus was to begin assembling a coordinated set of agency comments in anticipation of the coming legislative hearings.

Back home, adoption of the compact by four of the New England states had created a good reason for the New England Council to reexamine its previous stand. The council should not take a position counter to a majority of the states, Natural Resources Manager Francis E. Robinson argued. The newly consolidated Natural Resources Committee agreed, and the council was free to put its full weight behind the compact proposal. Special Assistant Donald Whitehead, on loan to the council from the insurance industry, was assigned to help Robinson with this undertaking.

The first test came in Washington, March 30–31, when the House Public Works Committee held a public hearing on the McCormack bill. This was no session for the faint-hearted, for opposition to the compact was expressed by virtually every federal agency, not the least of which was the Bureau of the Budget. As council executive vice-president Caverly reported later, the opposition even extended to many individual members of the committee.

NRC and its supporters certainly put their best foot forward. William S. Wise presented the NRC position in detail. The New England Council had enthusiastically supported the compact after its adoption by four states, and now Francis E. Robinson expressed the council's keen interest. Support came from such diverse outside groups as the League of Women Voters, the U.S. Chamber of Commerce, and the Water Research Foundation (proponent of a similar approach on the Delaware).

Governor Del Sesto of Rhode Island put in what he thought would be a token appearance on behalf of NEGC and was astonished by the hostility he encountered. Even House minority leader John W. McCormack drew bipartisan fire from the committee.

Although formal hearings on the Senate side had not yet been scheduled, the Senate Public Works Committee had at least agreed to solicit advance agency views on the compact bill. The long-awaited statement of March 31, 1960 from the Department of Justice did the compact cause little good. Justice traced the relationships between NRC and IACWR in considerable detail, implying that the New England group had deliberately failed to heed the earlier warnings about a course of action then and still deemed unconstitutional. What Justice considered as reasonable alternatives had been rejected summarily by NRC. The committee was advised in no uncertain terms not to acquiesce to the proposed joint federal-state arrangement.

At the April 14 NRC meeting in Boston, compact committee chairman Wise admitted that the prospects for passage in this session of Congress did not look bright. The matter of the unanimously negative reports from the federal agencies came in for considerable criticism, but the federal members observed that NRC had been fully forewarned. It was concluded, however, that these agency statements had foreclosed any further discussion on the compact by federal NRC members and that NRC per se could now take no official action. The states were, more than ever, clearly on their own.

The possibility then arose of a compromise to meet congressional objections, but the states had little latitude for adjustment because of the specific language of their own compact measures.

Despite the unfavorable prognosis for passage, NRC and the New England Council continued to work hard for the compact, receiving unusual support and cooperation from Congressman McCormack in this regard. As House minority leader, he was privy to much inside information and a figure few members of the House wished to offend.

On June 10, 1960 a so-called clean bill, H.R. 12467, was reported favorably by the House Public Works Committee containing only minor, face-saving amendments. This action was a tribute to New England persistance, but more particularly to Congressman McCormack's skillful leadership.

Aside from the House minority leader's influence, however, the second most decisive factor was a persuasive memorandum[63] prepared by the Public Works Committee staff, which was circulated widely within the halls of Congress. It rejected most of the objections raised by the federal agencies point by point, observing that the only credible reason for such severe and united opposition must be the dictatorial policy being exercised by the Bureau of the Budget.

The memorandum began with the very real problem of agency coordination in water resources development, noting that an already difficult problem is compounded manyfold when both state and federal agencies are involved. On the constitutional question, the memorandum noted that the New England proposal was not without precedent. Under the Potomac Basin Compact, for example, three state and three federal members, each with full voting rights, had operated an interstate compact commission without appreciable difficulty since 1939. The memorandum stressed the signatory action already accomplished by a majority of the states. "The burden of proof, and a heavy one, is upon those who urge that this compact be discarded and that the states be required to start all over again," it opined.

With regard to the recommended system of nonvoting federal representation, the committee staff observed tartly: "The arbitrary position of the Federal agencies on this question is not supported by any demonstration that the Federal observer system has proven a great success." Because of the fear of some federal agencies that a voting representative might arrogate to himself the powers of his organization, the committee staff observed that a nonvoting member is much more likely to be a nonuseful member. By giving him the power to a vote that would not bind his respective agency, this objection could be easily overcome.

Although it observed that there is no "royal road to effective coordination," the memorandum concluded by reminding the federal agencies of the recommendations of the Presidential Advisory Committee on Water Resources Policy, transmitted to the Congress by President Eisenhower on January 17, 1956. The committee staff noted that the proposed compact seemed to conform squarely with the Administration's own intentions of providing the states and local water resources agencies more adequate voice in the planning and development of projects.

In a letter conveying the good news of favorable committee action to the New England Council's executive vice-president, McCormack could not resist tweaking the elephant's tail. Observing that ten of the fourteen Republican members of the committee had filed dissenting minority views, McCormack stated: "It looks to me as though the Republicans on the Committee are going to try and make this bill, which means so much to New England, a Party fight. Frankly, this is rather strange to me, particularly in view of the fact a number of fine business organizations in New England, among which are included the Greater Boston Chamber of Commerce and the New England Council, strongly support the bill that I have introduced."[64]

Compact committee chairman Wise told the June NRC meeting in Boston that the prospects for enactment were now much more encouraging than they had been a few months earlier. He advised the June session of NEGC likewise, adding that he and his NRC associates intended to try to persuade the Senate Public Works Committee to waive further public hearings and support the House measure without further amendment.

By the end of August, all but the most evangelical of proponents had given up. Despite committee approval, the measure had not yet reached the floor of the House and almost certainly would not before Congress adjourned.

At the September 14 NRC meeting in Berlin, Connecticut, compact committee chairman Wise sounded the official death knell—H.R. 12467 had expired with the 86th Congress.

As Hurricane Donna blew her way into New England, NRC's new chairman, William S. Wise assembled his group in Berlin. It was not entirely an ill wind, however, for passage of Public Law 86-645, the Flood Control Act of 1960, had assured New England of some twenty-nine new civil works projects, of which twelve were for flood control. Lying in the eye of its own compact hurricane, NRC could afford to spend some time on other activities before round three of the compact battle would begin.

Internal problems were continuing to plague NRC. Despite valiant efforts by liaison members, and occasionally productive committee projects, the vast superstructure of subcommittees authorized previously was just not working. A second vice-chairman was authorized by NRC for the new fiscal year whose sole job would be to keep the various committees functioning properly. In addition, there were discussions of NRC's objectives and achievements and of suggested internal improvements.

At the November meeting in Augusta, Maine, NRC readdressed itself to the question of an interstate compact. It was agreed that the next moves would be on two fronts: the introduction of enabling legislation in the two remaining states and the reintroduction of H.R. 12467 into the 87th Congress.

1961

By the January meeting in Providence, Rhode Island, chairman Wise was able to report the introduction of a new compact bill by Congressman McCormack (H.R. 30). Wise had also attended in the interim the Interstate Conference on Water Pollution held in Washington, December 12-14, 1960, where there was increasing sentiment on behalf of the New England compact. Prospects for acceptance of the compact now appeared more promising than ever before.

By February, two more congressional bills had appeared: H.R. 2437 by Congressman Merrow of New Hampshire, and S. 374 by Senator Bush of Connecticut.

Two closely related national events also occupied a share of NRC's attention. Senate Report 29 had been released on January 30, conveying the findings of the Senate Select Committee on National Water Resources. Among its five principal recommendations was a ten-year, $50 million program of grants to the states for comprehensive water development and management planning. These plans would be prepared jointly by the federal and state governments much as New England had been proposing to do within its own region for the past several years.

Of more direct significance was the report of the Delaware River Basin Advisory Committee, which was recommending a federal-state compact for the Delaware with substantially greater powers than those proposed for the New England compact commission. Capable lawyers had reportedly viewed the proposed Delaware compact favorably on the question of constitutionality. The prospect of support at least in principle, and possibly in practice, from the powerful congressional delegations of New York, Pennsylvania, and New Jersey and the politically strong mayors of New York and Philadelphia, made the 87th Congress look like an entirely different proposition to New England compact proponents.

NRC was advised of continued support for the compact proposal from the New England Council, which had achieved important new capabilities nationally with the appointment of its first Washington counsel, Charles W. Colson, in January. Familiar with the compact issue from his prior service on Senator Saltonstall's staff, Colson went right to work on the legislation. The first step was the preparation for distribution to the New England delegation of a point-by-point brief supporting the compact legislation, including persuasive answers to the objections raised the previous year. Council Economic Research Director Rudolph W. Hardy took the task in hand, and by early spring he and Colson had a supporting brief in the hands of the New England delegation.

On the legislative side, however, further difficulties had arisen. McCormack's bill, by virtue of its identity with the previous year's committee report, was construed as not requiring further public hearings. Yet his adamant position against changing even a comma in his bill had angered the Republican members of the Public Works Committee to the extent that the measure was substantially bogged down. In some observers' eyes, it might not be acted on at all during the first session.

An additional conflagration had also broken out — regrettably, in the heart of the New England delegation itself. Early in June Senator Aiken of Vermont questioned the need for any more interstate compacts in New England, observing that Vermont had always gotten a bad deal on such joint projects. He was particularly fearful of federal domination. Regrettably, Aiken's stature and reputation had already swayed his colleague, Sen. Winston Prouty, and even New Hampshire's Sen. Norris Cotton, a cosponsor of the compact legislation, was reported to be wavering in his support.

Further evidence of defection was also apparent back home. Despite the early optimism, Vermont's enabling legislation had to be withdrawn because of Aiken's effective arguments and the opposition to the consent legislation in Washington. In Maine, it had been deemed prudent not to enter the bill at all!

By midsummer, however, conditions had taken a turn for the better. Joint Resolution 225 had been approved by the House providing for congressional consent to the Delaware River Basin Compact. This gave New England a strong precedent to work from. Furthermore, the New England compact proposal had picked up powerful national support from resolutions of the National Governors Conference on June 28 and the National Association of Attorneys General on June 14.

On July 12 the House Public Works Committee reported the compact bill favorably. One week later, as the result of adept maneuvering by McCormack, a rule was granted permitting the measure to be brought to the floor.

The Associated Press story from Washington reported two hours of limited but spirited debate during the August 2 House session. The Republican leadership continued to press for elimination of the federal voting powers, arguing that these provisions would give the federal agencies absolute control over the compact commission. Not so, said McCormack, as such amendments were rejected, commenting that it was "sickening to see the limited and narrow-thinking argument"[65] against the proposal. In the final hour, however, a measure of retribution was won by the Republican side. The provision for a $50,000 annual federal appropriation was eliminated in the approved bill.

With the New England compact bill now safely past the House, what had appeared to be only peripheral problems in the senatorial delegation now assumed substantial proportions. A way had to be found to persuade Senator Aiken that New England was not being sold down the river. The best approach seemed to be by way of the New England Governors Conference. At its September meeting, a formal request was made by the New England senators for the governors' current position on the pending compact. This was to be a critical moment, Colson advised in a confidential memorandum, for a firm and unanimous expression of interest in the compact would assist the cause immeasurably in Washington.

The minutes of the meeting state tactfully that "differences of opinion were expressed."[66] Four of the governors reaffirmed their favorable stance, but the governors of Maine and Vermont abstained because their legislatures had not yet acted favorably on the compact. Although the split was virtually unavoidable, it must be observed that the NRC-proposed legislation was never regarded as a "must" item at any time by the governors.

Fortified by policy divisions back home, the senatorial opponents of the compact had their convictions strengthened by another apparent example of proposed federalization of water resources. President Kennedy, on July 13, had transmitted legislation to the Congress designated as the Water Resources Planning Act of 1961. Although the river basin commissions to be established under this bill would consist of both federal and state members, all appointments would be made by the President. So serious did the consequences appear that the Council of State Governments distributed a memorandum dated July 31, 1961 to all governors, state legislative leaders, attorneys general, commissioners on interstate cooperation, and interstate water conference members that observed that the measure, if enacted, would affect radically the respective federal and state roles in the water and related land resources field.

Optimists, however, saw in the emerging national issues a ray of hope in meeting Vermont objectives. Perhaps Sen. Aiken could be encouraged to oppose the Administration's water resources planning bill, thereby placing the New England compact in the perspective of a preferable alternative. And if an amendment could be obtained requiring the existence of a regional compact prior to the establishment of a river basin planning commission, then New England might still win its case.

1962

In the meantime, the Senate Public Works Committee had proceeded to the point of soliciting views from the various federal agencies. The statement of January 11, by Secretary of the Interior Stewart P. Udall to committee Chairman Chavez is noteworthy for its moderation of the previous federal agency positions. Udall, understandably, preferred to back the Administration's pending water resources planning act. He did, however, state that the compact commission visualized by NRC "can be adapted to performing the necessary planning work in a manner consistent with the planning program envisaged in the Water Resources Planning Act."[67]

By mid-March, Washington counsel Colson privately expressed his concern over the lack of substantive action on the Senate side and enlisted the help of the New England Senators Conference, which tried to persuade the Judiciary Committee to discharge the House-passed H.R. 30 to the Public Works Committee, where S. 374 was already lodged and which had reviewed and reported the Delaware River Basin Compact legislation. Meanwhile, the state NRC members and the New England Council's Natural Resources Committee wrote the chairmen of both committees urging prompt action on the bills.

On the heels of Walter White's discouraging note to his fellow NRC state members late in July stating that the compact would probably not see the light of day came the exhilarating news that the Senate Judiciary Subcommittee under Sen. Thomas Dodd of Connecticut, an avowed compact supporter, had at long last scheduled formal hearings.

Although the New England proponents put on a brave performance at the September 18 hearing, it was the Vermont senators who carried the day. As the UPI story the next day stated, "Vermont Senators Flay Plan for New England Resources."[68] Aiken was quoted as characterizing the compact as "weasel worded" and "the worst that has ever been proposed." Whoever prepared the compact, Aiken stated, must have been "an expert experienced in the art of spreading confusion and obfuscation," and he damned the New England Council with faint praise for "avidly promoting" the compact legislation.

Although the subcommittee did ultimately report the measure favorably, the full committee did not have time to act before Congress adjourned. Had it done so, however, Vermont's two senators were reported by the *Concord* (New Hampshire) *Monitor* as already having "mounted guard,"[69] fearing an attempt to put through the compact in the Senate during the closing days as an amendment to another bill. For the second straight session, New England's compact aspirations had been dashed.

Faced with so discouraging a turn of events, New Hampshire representative Walter White wrote his fellow New Englanders on October 16, 1962 that the time had come to appraise candidly the future of NRC. While not disagreeing with this conclusion, Connecticut's Wise was still hopeful, stating that it would be a mistake to entertain NRC's demise just because the compact legislation was lost in the closing shuffle of Congress.

However, what had been troubling White for more than a year was NRC's inability to maintain what its own charter called for—an up-to-date and dynamic program for the development of the region's resources. Several proposals had been entertained for updating the previous NENYIAC report, including a pilot undertaking for the Connecticut River Basin. The federal agencies, however, had observed that it would cost virtually as much to do a single river basin as to do the region as a whole.

Nevertheless, NRC had taken the White proposal seriously, and its executive committee investigated the scope and costs of a regional resources planning program. It found that, despite the preexisting NENYIAC data, more than $24 million would be required to replan New England's water and related land resources comprehensively, of which more than $4 million would be required by the states. Because of sensitive agency toes, the committee avoided any consideration of project scheduling other than to estimate that some $1 billion in project funds would be required to carry out the resultant plans.

From 1961 to 1962 the NRC had continued to tinker with its internal machinery, hoping to arrive at some realistic and productive meeting and committee structure. Also, after prolonged discussion, three study items of top regional priority had been identified: benefit-cost analysis, flood plain zoning, and multiple use of water supply reservoirs.

At the September NRC meeting in Berlin, Connecticut, the Corps of Engineers discussed the details of its recently authorized comprehensive study of the Connecticut River Basin. This came somewhat as a surprise to many NRC participants in the light of previous discussions over NRC's possible role in comprehensive planning. The authorization for the Corps study seemed to coincide closely with the figures for the Connecticut basin contained in the previous report of NRC's own ad hoc committee!

In other developments, the New England Council's director of interstate relations, Dr. Peter C. Janetos, doubling as secretary of both NRC and the Organization of New England–New York Planning Directors, suggested that a common cause existed between the organizations in the resources planning field. Closer cooperation was urged. It was agreed that Vice-Chairman Abelson of Interior would pursue this possibility further on behalf of NRC.

At the November meeting, the decision was made to have the compact reintroduced into Congress and to furnish the New England governors a complete report on the compact issue at their March 1963 meeting.

1963

By the beginning of 1963 a new note had crept into NRC discussions. Might not the states be just as well off under the proposed modifications to the pending national water resources planning act?

The state water resources directors were quite cognizant of this possibility through their association with the annual Interstate Conference on Water Problems. However, after prolonged discussion, the home-grown compact still seemed the most desirable solution. Sen. Aiken was consulted on possible modifications that might make the compact more acceptable to him. Executive Vice-President Gardner Caverly of the council, a former Vermont businessman, was selected as the best contact with Aiken. Colson, New England Council representative in Washington, found that his suspicions of unrelenting opposition were confirmed. As Caverly reported to the March meeting of NRC, the senator advised that he would continue to vote against it.

By this time, a compact bill had been entered in the House under joint sponsorship of Representatives Daddario of Connecticut and Conte of Massachusetts. Senator Dodd had expressed similar intentions but was held off on the grounds that a northern New England sponsor, such as New Hampshire's Senator Cotton, would engender the least resistance from Senator Aiken. Closer to the Washington scene than any of the others, Colson continued to advocate some alternative course of action to the compact if at all possible.

During the May 1963 meeting of NRC held in Norwich, Vermont, matters began to come to a head. Considerable time was spent on the provisions of S. 1111, the new version of the Water Resources Planning Act just introduced by Senator Anderson, who was reported to have incorporated into it many suggestions from the Interstate Conference on Water Problems. Consequently, the measure now had the complete backing of the Council of State Governments, the National Governors Conference, the General Assembly of the States, and the major state water resources administrators.

White of New Hampshire, historically the earliest of the New England compact advocates, pressed immediately for endorsement of S. 1111 by NRC. The states had been given every consideration, he argued, and unwillingness to support the measure might be construed as bad faith on their part. His feeling was shared by Thieme of Vermont.

Connecticut's Wise, however, found it difficult to support the national legislation when New England's own compact proposal was still pending. He reported that some eight compact bills had now been introduced into the 88th Congress, and any deviation from this established course might prove damaging to NRC's congressional support.

So divided were the opinions that formal action was postponed until the June meeting in Portsmouth, New Hampshire, and advance warning was given that the matter would be put to a vote. The decision was that no action would be taken on S. 1111 at that time. The Council of State Governments was consequently informed by formal resolution that NRC was still vitally interested in its own compact proposal.

1964

By January 1964, even Wise's loyalties to the compact cause were beginning to waver. As he told NRC at its executive session in Boston on January 9, the various compact bills were relegated to a virtually inactive status by the growing support for S. 1111, and Wise suggested that an amendment to it could insure that NRC became the river basin agency for New England should the national legislation be approved.

By the April meeting, Wise was convinced that the compact was a dead issue and suggested that it be dropped from further discussion. S. 1111 had already been reported favorably by the Senate, and hearings had recently been held by the House Interior and Insular Affairs Committee with the prognosis for passage currently favorable. Massachusetts representative Foster suggested that NRC pursue passage of this bill and recommend a specific course of action for the New England governors to follow. With the help of the Council's interstate relations director, Edwin Webber, a summary statement was prepared so that the governors could be briefed individually prior to their May session.

By the June NRC meeting in Salisbury, Connecticut, there was substantial agreement among NRC members on the desirability of formal support for S. 1111. It was reported that the governors were receptive and would take official action at their September meeting. Consequently, on June 18, NRC adopted itself modestly as the "logical agency to be authorized to function as the river basin water resources commission in New England under Title II of Senate Bill 1111."[70]

Toward the latter part of August, prospects of congressional enactment of S. 1111 grew appreciably better. Both President Johnson and Speaker McCormack were reported as taking a personal interest in the water resources planning measure. Regrettably, however, the timetable was just too limited, and Congress adjourned without final action on the bill. But virtual assurance was given that S. 1111 would be among the first items considered by the 89th Congress in January.

Accustomed by now to such adversities, NRC assembled in Cranston, Rhode Island, in September to tend to current business. In a further attempt to streamline its operations, it was determined that meetings would be held at quarterly intervals and all of the various nonfunctioning standing subcommittees would be abolished.

Modeled on the approach of the New England Interstate Water Pollution Control Commission, a single Technical Advisory Committee was to be created made up of the chairmen of the previous work groups, and a special liaison was established between this body and the newly created state Water Resources Research Centers.

An intriguing alternative to S. 1111 arose in the form of the interstate compact for regional planning that was being encouraged by the New England governors. Since regional planning fell within the prerogatives of the Housing and Home Financing Agency—later the Department of Housing and Urban Development—the compact would not require specific sanction by Congress. Why couldn't a marriage be arranged between the state planners' proposal and the NRC-advocated compact, thus accomplishing indirectly what NRC could not win directly? To NRC, this seemed worth a try.

A secure bridge was at hand in the form of Edwin Webber of the council's interstate relations staff, who was serving both as secretary of NRC and as a participant in the planning discussions. Through his good offices, special subcommittees of both groups were appointed to pursue the matter further. Two joint sessions were held prior to the scheduled New England Governors Conference in Boston on December 9. As a result, a New England–wide planning program was presented personally to the governors by Vermont Governor Hoff and endorsed for submission to the respective legislatures.

On December 10, Chief Counsel Mitchell Wendell of the Council of State Governments, original drafter of the NRC proposal, found himself retreading historic ground as he arrived in Boston to try his hand at another New England compact. NRC's interest in the regional planning compact, however, proved to be largely peripheral. Once a united and coordinated front had been presented to the governors, and resource planning determined to be the primary province of NRC (or its future equivalent), NRC's attention became refocused on the national water resources planning legislation.

1965

At the January meeting of NRC in Boston, chairman Foster reported reintroduction of the measure into Congress as H.R. 1111 and S. 21 and recommended its formal endorsement by NRC. By subsequent resolution, NRC reaffirmed its previous support of the Water Resources Planning Act but tempered its earlier statement by suggesting only that NRC become "an important part" of any regional commission within the northeast region—not necessarily the actual agency.

By the next meeting in Berlin, Connecticut in April, action on the legislative front was well underway. The interstate planning compact had been introduced into the six state legislatures, and favorable hearings had been held in all. At the Washington level, the water resources planning legislation had cleared both branches of Congress and was awaiting conference committee resolution of the few differences remaining. If NRC was to be in the forefront of future action under this act, it clearly must move promptly. A special committee was, therefore, established under Thomas J. Rouner, former NRC Evaluation Standards Committee chairman, and assigned the task of reviewing S. 21 and recommending appropriate action to NRC at its next meeting.

Rouner was well chosen. A vice-president of the New England Power Company, he had been an active participant in both NENYIAC and NRC affairs. Experienced in water resources and highly respected by both state and federal agency personnel alike, he could be counted upon to provide the neutrality a delicate mission of this sort required.

The Rouner committee reported back to NRC at its meeting in Whitefield, New Hampshire in June. A detailed proposal was presented, under which New England would have the first river basin commission in the nation. Once this was accomplished, however, NRC would cease to exist.

Drastic though its recommendations were, the Rouner report generated surprisingly little discussion. As if relieved to settle upon a definitive course of action at long last, NRC voted unanimously to accept the report. In further action, the report was to be brought to the attention of both NEGC and the newly established Water Resources Council (WRC), successor to IACWR in Washington and the federal administrator of the new Water Resources Planning Program.

At the September meeting in Bethel, Maine, chairman Foster reported his mission at least partially accomplished. He had met with the governors in Springfield, Massachusetts on September 18, and they had not only agreed to request formally the establishment of a New England River Basin commission, but also to pledge a sum of not less than $5,000 per state to cover the nonfederal share of the commission budget for the balance of the fiscal year. This was not only support, but actual cash on the barrelhead. With these encouraging developments in mind, NRC adopted a resolution making preparations for phasing itself out.

On October 14 New England received the welcome news that WRC had officially approved the governors' request. The region received high marks from the council for being the first in the country to propose a river basin commission. A New York member had been added and the title pluralized to New England River Basins Commission. Otherwise, the approved version was the same as the one New England had submitted originally. Establishment now seemed a mere formality to be accomplished at any time by the President.

1966

By the time of the meeting in Boston in January the winds of change still appeared to be soft and favorable. State designees had been appointed to the proposed river basin commission, and Austin H. Wilkins of Maine had been elected temporary vice-chairman. Wilkins explained to NRC and its invited guests the state members' preliminary thoughts on commission location, size, and composition, and noted that his group was submitting to Secretary Udall, chairman of the Water Resources Council, a slate of candidates for the position of chairman.

The NRC special committee on deactivation made a number of recommendations concerning the phasing out of NRC and the composition of the new commission. Once it had been officially established, the records of NRC and any pertinent information in other agency files should be turned over to the new organization. As a matter of protocol, NRC Vice-Chairman Wilkins was to advise the New England governors of the decision to deactivate. Again through Wilkins's good offices, the eligible interstate agencies would be contacted for official representatives. Each federal representative, in turn, was to request his office to appoint a formal designee to the commission in the near future.

With dissolution so strongly in the air, the next meeting of NRC was set tentatively for April. Deactivation would automatically take place if the commission were to be established in the interim.

As NRC met in Boston on April 14 for an abbreviated session, establishment of the new commission was expected momentarily—in the eyes of one knowledgeable Washington observer, at least by the May 16 meeting of NEGC. NRC was assured that the delays were due only to minor difficulties inherent in the formation of new commissions. A presidential executive order had allegedly been drafted and was under review at the White House. Most of the remaining speculation centered on who was to be the federal chairman.

April stretched into June and June into September, and there was still no action in sight on the proposed New England commission. Soundings made in various Washington channels revealed a variety of excuses, from budgetary problems to staffing difficulties to procedural disagreements to an international jurisdictional problem on the Canadian border. Most disturbing of all, the Administration was reportedly "hung up" on the selection of a chairman.

By early August the New England Division engineer, Col. Remi Renier, current chairman of NRC, felt he had waited long enough. He directed his staff to set up an NRC meeting for September.

In the interim, he learned that WRC director, Henry B. Caulfield, Jr., had contacted vice-chairman Wilkins, confirming the budgetary problem faced by the new commission. Caulfield had inquired whether the governors would be willing to raise their pledges to $10,000 per state, or the equivalent of 60 percent of the proposed first-year budget. This was in sharp contrast to the two-thirds federal, one-third state cost sharing anticipated earlier by the state designees, and the dollar-for-dollar matching they considered the absolute maximum. Following consultation with Governors Conference chairman Reed and others, it was agreed that the matter would be brought before the September 18 meeting of the governors for their consideration.

At their session in Springfield, Massachusetts, the governors discussed the lower level of federal contributions than had been predicted earlier. However, it was pointed out that the sum of $10,000 per state was still consistent with the earlier action of the governors, in which the $5,000 per state had been pledged for the half of the fiscal year then remaining. A pragmatic resolution was adopted committing each governor to seek from his legislature the sum of $10,000 — "if we are elected"![71]

At NRC's September meeting in Rye Beach, New Hampshire, WRC director Caulfield was placed squarely on the spot. He responded with details that made it clear that there was much more to the establishment of a New England commission than NRC had suspected.

NRC then addressed itself to its future activities. It was felt that NRC could profitably direct its efforts toward the preparation of a draft program for the new commission. This would save valuable time and would hardly be duplicative, since all but the federal chairman had now been appointed. NRC chairman Renier of the Corps set about the task.

In accordance with the understanding reached at the recent NRC meeting, chairman Renier addressed an inquiry to all NRC members on October 3 soliciting recommendations for the proposed commission program. His intention was to use these to prepare a framework for its operations that would include a definition of its mission, a ten-year plan inclusive of all foreseeable federal planning, and a specific plan for the most immediate five-year period.

Renier had suspicions that the states were losing interest, and these were well founded, as was proved in his discussions with representatives of four of the states following a meeting of the Connecticut River Coordinating Committee in late October. Renier told them he was fully prepared to proceed with the definition of a program and budget for the proposed river basins commission in his capacity as NRC chairman. The four recommended strongly against any further work on the budget by the Corps, fearing that NERBC would become just another commission dominated by federal agencies, and not the clear-cut partnership envisioned when the New England governors were first approached. They were discouraged with what they considered an about-face on budget support by Congress, which would make the desired state participation even more difficult to secure. Finally, the state representatives were provoked more than words could express by the interminable delay in the commission's establishment, feeling somehow that they were caught in a policy squeeze between the Corps and the federal WRC.

1967

On January 10 NRC assembled again in Boston. The principal item of business was a draft entitled *Potential Program for the New England River Basins Commission,* synthesized by the Corps from comments received from NRC members. The consensus was that the recommended program, though broad, was acceptable since it could only be general in nature at this time. Following suggested revisions, it was to be placed in the hands of the New England governors and WRC in the near future for information purposes. With commission establishment believed just around the corner, the matter of the next meeting was left to the discretion of the chairman and vice-chairman.

On March 6 an already discouraged New England group read a newspaper announcement they could hardly believe—the Pacific Northwest River Basins Commission had just become the first river basin commission in the country. In acknowledging the news, vice-chairman Wilkins could not help observing to Colonel Renier that he continued to receive questions he could not answer on why the New England commission had not yet been established.

By June, however, word had leaked that some action on the New England commission was imminent, and the wheels were set in motion for the long-awaited transition from NRC to NERBC. A September meeting was tentatively scheduled to greet the new federal chairman and formally hand over the reins of office.

In reporting that the New England River Basins Commission had finally been established on September 6, Washington reporter William H. Young of the *Providence Journal-Bulletin* observed wryly that the commission had "bobbed to the surface" nearly two years after WRC's favorable recommendation had "disappeared into the opaque waters of the Washington bureaucracy."[72] When the action finally came, New England was the fourth, not the first, such commission in the country.

Simultaneously with the establishment of the new commission came the announcement that R. Frank Gregg, a career Washington conservationist and a Coloradoan by background, would become its first chairman. The *Providence Journal-Bulletin* in an editorial observed that it was unfortunate a New Englander could not have been found who was acceptable to all the Democratic powers in New England. Yet it viewed Gregg's appointment as a possible blessing because of his fine qualifications, his lack of preconceptions about New England, and his absence of ties to any existing groups or interests in the region.

The 67th and final meeting the NRC took place in Waltham, Massachusetts on October 6, 1967. Though long planned, it had its awkward moments. The new chairman of NERBC was introduced warmly by Colonel Renier and spoke briefly of his admiration and aspirations for New England.

NRC then directed its attention to its own demise. This created considerable difficulty, since no one could recall precedent for bureaucratic suicide. Moreover, the original charter-granting agency, IACWR, had gone out of business. A formal vote was finally taken, authorizing chairman Renier to seek whatever action was necessary from WRC, the only likely successor to IACWR. Vice-chairman Wilkins was also empowered to draw up a memorandum of understanding for the governors' signature approving NRC's dissolution.

NRC had finally closed the doors on its ten-year history of activity in New England.

CHAPTER FOUR

The New England River Basins Commission

1967

The organizational meeting of the fledgling New England River Basins Commission (NERBC) took place on October 16, 1967 in the John F. Kennedy Federal Office Building in Boston. In anticipation of a sizeable volume of business, a two-day session had been scheduled. Chairman Gregg's baptism by fire had been preceded by efforts to get to know some of the principal players in the course of what had become a regular commute from Washington. The New England Division of the Corps had been genuinely welcoming, first offering NERBC space at its offices in Waltham and then making available its real estate specialists to help find suitable quarters in Boston. Colonel Renier, the division engineer of the Corps and the current chairman of NRC, had issued instructions to his staff that every courtesy should be extended the new agency. Lacking even a secretary, Gregg had leaned on the Fish and Wildlife Service to help him to prepare materials for the first meeting.

The room was full that auspicious afternoon. Every federal and state agency had an official representative on hand. Of the interstate members, only the Atlantic States Marine Fisheries Commission and the Merrimack River Valley Flood Control Commission failed to be represented officially.

The executive director of the Water Resources Council (WRC), Henry P. Caulfield, Jr., had flown in from Washington to attend. The commission would be considered organized when it had held an organizing meeting and, by consensus, had agreed to proceed with its work, he said. He outlined the functions of the commission: 1) to serve as the principal agency for the coordination of planning; 2) to prepare and keep up to date a comprehensive, coordinated, joint plan; 3) to recommend priorities for planning and construction and 4) to undertake such studies as might be necessary. Caulfield was quick to note the differences between NERBC and NRC. NERBC's planning and coordination mandates would include the full gamut of federal, state, interstate, local and nongovernmental activities. It was truly a federal-state organization as reflected in staff, budget support, and the unique provision of a coequal vote if consensus could not be achieved. Unlike NRC, the commission would have a specific mission. As unprecedented ventures in the long history of national water policy, the river basin commissions would be expected to develop new approaches and new relationships.

At the chairman's request, Colonel Renier reported that the 67th meeting of NRC, held on October 6, would be its last. The committee had affirmed resolutions for its own demise. Letters would be dispatched to WRC and the New England Governors Conference (NEGC) seeking permission to dissolve, since NRC had been established by charter, and the committee was prepared to turn over its official records at this time. In its first official actions, NERBC resolved to accept the data, records, and files and to express its deep appreciation to NRC for its interim coordination and leadership.

Certain parliamentary matters were then taken up. In the absence of a set of by-laws, temporary operating procedures had to be agreed upon. The most sensitive was the definition of the term *consensus*, the statutory basis of the river basin commission's actions. Caulfield defined *consensus* as the lack of objection, as contrasted with unanimous action, which would require a positive sign by all members. Thus, an official representative could approve, object, abstain, or submit a reservation in writing for the record. Given a stand-off among state and federal members, the law provided that the chairman should set forth the federal position and the vice-chairman the state position. Caulfield pointed out that Public Law 89–80 was drafted this way so as to encourage alternatives to be set forth in order to reach consensus. In many respects, the consensus-building features of the act were its most central provisions.

A question arose about the role of the interstate agencies. They would be included among state members for purpose of determining a quorum, the commission decided, and allowed to vote on all matters except the election of a vice-chairman.

In the matter of budget, Caulfield advised the commission that the federal share was limited to 50 percent of the costs in addition to the expenses of the chairman. WRC had established a ceiling of $200,000 in fiscal year (FY) 1968 federal contributions for any given commission. Maine's Austin Wilkins reminded the group of the New England governors' earlier pledge of $10,000 per state, but neither New Hampshire nor Rhode Island currently was in a position to fulfill that commitment. However, since a number of states reported funds in hand and only $84,000 would be required for the balance of the year, the commission adopted a tentative budget for FY 1968, a revision of NRC's earlier estimates.

Before doing so, the state members elected a permanent vice-chairman for, under the law, most NERBC administrative actions would require the concurrence of the state officer. Austin Wilkins of Maine was the unanimous choice.

After personnel matters had been discussed, arrangements were made for an ad hoc committee to develop procedures and priorities for the commission's work program.

In final action, the officers of NERBC were instructed to notify WRC and the governors that "the commission has perfected its organization and is ready to assume the obligations, duties, and responsibilities imposed upon it by law and the Executive Order aforesaid."[73]

In retrospect, the first meeting had been a good one. Chairman Gregg had shown himself to be organized, firm, but diplomatic. The states had modified his draft resolutions to assert their independence, but not to be obstructionary. The consensus process had been employed and it had worked. It was clear that there would be no split between federal and state interests. Differences would more than likely occur, but the alliances and confrontations were apt to be individual. The long and painful heritage of NENYIAC and NRC had demonstrated one fact: state and federal representatives knew how to work together.

Within a month, NERBC was back together for its second official meeting, again in Boston. The principal order of business was a briefing on three major New England programs: the North Atlantic Region (NAR) and the Northeast Water Supply (NEWS) studies of the Corps of Engineers, the New England Comprehensive Water Quality Study of the Federal Water Pollution Control Administration (FWPCA) and the program of the New England Regional Commission (NERCOM). Chairman Gregg, from his travels throughout the region and his discussions with executive director Caulfield of WRC, knew that his commission would not succeed unless it asserted a role for itself in these activities.

Gregg reported to the commission his brief appearance before the November 16 meeting of NEGC. The governors were interested in planning, he said, but only as a step to responsible action. They were frustrated by the large number of programs and organizations in the water resources field and hoped for some recommendations from the commission for strengthening and consolidating these functions within the region. Vermont's Gov. Philip Hoff, the current conference chairman, had subsequently written him urging that the commission tackle specific projects with regional implications, such as the downstream thermal effects of the Vermont Yankee nuclear plant.

Gregg had been able to attend several meetings and consult with the vice-chairman in Maine since the last meeting. Pursuant to the discussions in October and the interests of the governors, he had appointed Alfred E. Peloquin of the New England Interstate Water Pollution Control Commission (NEIWPCC) as chairman of an ad hoc committee on interstate agencies.

Burnham H. Dodge, chief of planning for the Corps of Engineers' North Atlantic Division (NAD), was then invited to describe the NAR water resources study, one of eighteen comprehensive studies projected by WRC to cover the needs of the United States by the year 2020. The study, covering an area from Virginia to Maine, was authorized by the River and Harbor Act of 1965, a significant expansion of the Corps' traditional mission. By way of contrast, the NEWS study, an outgrowth of the northeast drought, was concerned solely with water supply, reaching down to the region's municipalities and to the nearly 4,000 privately owned water supply utilities. Both studies were two to three years down the road and operated out of the Corps's NAD office in New York.

Chairman Gregg raised the question of NERBC's role in these two major regional studies. At the least, he said, the portions applicable to New England should be published as separate documents. He felt the chances of implementing action would be improved significantly if the region's own agency could be identified with the studies. Massachusetts's Robert L. Yasi and Connecticut's William S. Wise supported that position. As Wise observed, the states wanted the commission in the first place to help strengthen their voice in ongoing planning programs. Dodge objected to any fragmentation of the studies, and, in the absence of consensus, a committee was established to explore possible relationships and present recommendations at the next meeting.

The presentations by FWPCA went more smoothly. His agency was anxious to work closely with NERBC in coordinating regional water quality programs, FWPCA regional director Lester Klashman asserted. Staff members Walter Newman and Bartlett Hague described in detail their comprehensive water pollution control programs and pointed out certain subgranting provisions available to state or regional agencies.

John Linehan, Gregg's counterpart on NERCOM and the senior federal official for New England, then described the programs of his agency, one of a network of regional economic development commissions established under Title V of the Economic Development and Public Works Act of 1965. Linehan served as federal cochairman; the governors were the other members, and one governor served each year as state cochairman. Unlike NERBC, NERCOM had been funded at a level of $1 million for operations and another $5 million for grants. Its four priority areas for special study included transportation, pollution control, manpower development, and comprehensive plan development. The last area, in particular, suggested possibilities for fruitful collaboration. NERBC's Gregg noted that close cooperation might be achieved through joint staff meetings and coordinated hearings. He only wished that the charter and legislative history of the regional commissions were not so clearly tied to economic development as a social goal.

Before the meeting adjourned, Massachusetts's Yasi returned to the subject of Vermont Yankee. He had been approached by the chairman of the Vermont Water Resources Board for a meeting of downstream Connecticut River interests and felt that the NERBC would be the proper agency to bring the parties together. The commission raised no objection to discussing the matter at a future meeting if the concerned states so desired, feeling that it would be entirely appropriate for NERBC to serve as a forum for investigation and discussion. Colonel Renier reminded his colleagues of the interest of the governors in such matters and expressed the opinion that NERBC should try to be responsive at all times to their suggestions.

1968

By the January meeting of the commission in Boston the chairman could report a substantial measure of progress in acquainting the region with the new agency. He had touched base with business leaders, federal agencies, research specialists, and the League of Women Voters and had received a cordial welcome in all quarters. Chapman Stockford, the executive secretary of NEGC, was attending his first meeting in order to provide a firsthand report to his chairman, Vermont's Governor Hoff, on the matter of interstate agencies. Alfred E. Peloquin, spokesman for NEIWPCC, presented summary data for the region's six interstate compact agencies. All serve useful purposes, he reported, and should not be consolidated into NERBC operations.

NERBC's relationship with ongoing Corps studies returned to the agenda. The special committee agreed on continued Corps leadership but felt that separate volumes should be prepared for the New England portion of the region. From Paul Shore of the Federal Power Commission (FPC) came an unexpected objection. NERBC should reserve the right to publish a framework study of its own. New York's Francis W. Montanari agreed. The Commission should not commit itself irrevocably to the results of NAR or NEWS. In an eleven-point statement NERBC agreed to participate in both studies on a staff and committee basis in return for recognition as a principal but not exclusive point of contact for the region. The states would continue to represent their individual interests as NAR Coordinating Committee members. Chairman Gregg made it plain where he stood on the larger question of comprehensive studies. In the future, he observed, all multistate multiple-purpose water and related land resource planning investigations should be carried out in the name of and under the overall direction of NERBC. In cases where Congress directed a line agency to perform a limited or single-purpose study, NERBC should establish a coordinating committee under the chairmanship of the designated agency. The gauntlet laid down was a significant one, for many other comprehensive studies were in the wings. Gregg perceived correctly that his agency would never achieve its statutory goal of serving as the principal agency for coordination unless it also served as the principal agency for planning within the region.

Montanari of New York then turned to another issue. His agency took exception to the southern boundary of the NERBC region, which included only the Connecticut and Rhode Island portions of Long Island Sound. For the sake of hydrologic and ecological unity, the entire sound should be included. Montanari was encouraged to discuss the matter with the states directly affected and draft a resolution to be adopted by NERBC, endorsed by the governors, and then forwarded to WRC for action by the President.

Turning to the commission's substantive program, the chairman announced the creation of seven working committees, including committee assignments. Colonel Renier would head a special effort to identify priorities for new planning investigations. In accepting the assignment, Renier observed that the Commission should not get itself in a position where it can do nothing but long-range planning. Chairman Gregg noted the peculiar nature of the Comprehensive Coordinated Joint Plan (CCJP) requirement set forth in the Water Resources Planning Act. Any study deemed to be an element of the CCJP would not require further congressional authorization. Commission members spoke of several investigations they felt to be of particularly high priority. Beach erosion, comprehensive studies for both the Merrimack and the Blackstone, and a multipurpose study of the region's coastal resources were among those mentioned.

The chairman also shared with the commission his preliminary thoughts about staffing. He did not subscribe to the suggestions made by NRC, for technical experts in a number of fields would not necessarily provide the breadth of competence and experience he felt the commission would need.

By the March meeting in Boston the commission had settled down to a reasonably routine method of operation. State and federal agency reports were made a standard part of each session. The chairman would report on his activities; the vice-chairman on matters peculiar to the states.

Francis Montanari of New York then moved his resolution adding Long Island Sound to the commission region, observing that no other federal-state mechanism for planning coordination existed in this area. Although not objecting to the change if it was the desire of the states, the Corps's NAD argued by memorandum that the sound's economic orientation to New York City rather than New England could raise technical and political complications.

Neither had NAD given up on the matter of its NAR and NEWS studies. The price for separate volumes had risen to $70,000, and the commission did not think it wise to make such a request of Congress. Chairman Gregg indicated that NERBC would work closely with the Corps to obtain as much useful information as possible for New England, and it would still reserve the right to publish its own framework study.

A detailed memorandum prepared by Gregg on NERBC's goals and objectives, methods of organization, and operation was considered. He suggested four main areas of emphasis: policy and program coordination, plan development, community relationships, and administration. The staff and committee structure would mirror these elements. Gregg's concept was to assign the actual conduct of planning to the state and federal agencies responsible for such operations. The commission's role would be to develop the overall strategy, secure coordination and, upon occasion, assume staff leadership of a particularly broad investigation. The catch was, of course, full participation by other agencies. At their morning session the next day, members discussed the inadequacy of funding to permit effective participation. Gregg responded that the costs, upon occasion, could be reimbursable.

The special work group on priority planning reported to the commission at its May meeting in Boston. Priority was to be given to five program elements: 1) commission participation in the NAR and NEWS studies; 2) special projects relating to flood plains and the hazards associated with the safety of small dams; 3) production of a publication describing ongoing state and federal programs; 4) a comprehensive study of Narragansett Bay; and 5) review and comment on Federal Power Commission (FPC) licensing applications. Other possibilities were mentioned. The work group made it clear that it regarded the Narragansett Bay study as only the first in a series of comprehensive studies covering all of the region's major river basins.

The recommendations with respect to Narragansett Bay drew extensive discussion, for the Corps had been authorized to conduct such a study by a March 29 resolution of the Senate Public Works Committee. The tides of March had prompted Rhode Island's two senators to approach the Board of Engineers for Rivers and Harbors for a traditional, Corps-led investigation. It was Gregg's suggestion that the Corps utilize the March 29 resolution to authorize its participation in an NERBC comprehensive under provisions of the Water Resources Planning Act. If the Corps could focus its efforts on the immediate flood problems, then a heavy burden would be lifted from the proposed Commission study.

New Hampshire representative Roger J. Crowley, Jr. asked that a recess be called so that the states could caucus. The decisions on program emphasis had become central to the question of whether the states would be full and active planning partners. He returned to propose a resolution according first priority to the development of a CCJP for the region, whose program would include a framework study for overall guidance and comprehensive studies for major river basins and resource regions. The remainder of the action program would include special projects, such as cooperative subregional water pollution control programs, flood plain and small dam protection, and continuing responsibilities of a more general nature. Included in the last would be programs to review and recommend priorities, to encourage information exchange, and to report on power plant licensing and siting. Crowley's motion was seconded by the Commerce member, Ralph K. Kresge, and in a display of accord among state and federal members, NERBC's fiscal year 1969 action program was adopted without reservation.

But the trouble was not over. The commission still had to deal with the matter of Narragansett Bay. The state members from Maine and Rhode Island proposed a resolution recommending to the governors and WRC a comprehensive study of Narragansett Bay under NERBC leadership. This went right to the heart of the governors' intent in requesting establishment of the Commission, Gregg noted. Colonel Renier advised that the Corps could not relinquish its responsibility for the study unless the authorization was altered by responsible officials. Should the matter come to a vote, his agency would have to abstain. The matter was complicated still further by the activities of the

Task Group on Interagency Coordination, established by the National Council on Marine Resources and Engineering Development, which included the Corps of Engineers, the Department of Health, Education and Welfare, the Atomic Energy Commission, and the Department of the Interior. Its explicit purpose was to appraise the potential of a river basin commission as a medium for federal-state cooperation in planning. The Task Group had chosen Narragansett Bay as its pilot region.

A resolution was worked out and was approved by consensus with the Department of the Army abstaining. The chairman was instructed to meet promptly with Corps and WRC officials to achieve an amicable resolution of the policy issue.

Chairman Gregg then brought up the matter of future budgets. The carry-over of funds from FY 1968 would be gone by FY 1970. At that time, he estimated that state contributions would need to be raised from $10,000 to $20,000 each. Vice-chairman Wilkins reminded his colleagues of the past history of pledges from the governors to help get the commission started. Do the states now wish to go for full funding of the program or stay with the present $10,000 limit, he asked? The response was to move ahead. Perhaps the states' share should be apportioned more equitably, Vermont's Lemuel Peet observed, but he felt that the commission's three-year budget estimates should be put before the New England governors at their June 28 meeting in Stowe, Vermont.

Chairman Gregg could also report an agreement with NERCOM to draft a memorandum expressing joint goals and coordination procedures. The two agencies would co-locate in a building on Court Street just off Boston's Government Center.

But the most interesting news was Gregg's choice of his first two staff members. The staff director position had gone to Malcolm E. Graf, former director and chief engineer of the Massachusetts Water Resources Commission, a broad-gauged, self-educated individual with an encyclopedic knowledge of the region's water resources who was universally known, respected, and liked. His assistant would be Robert S. Restall, also an engineer, who had begun his career with the Corps of Engineers during the earlier NENYIAC study. His father was Wesley Restall, the New England Division project engineer who had been assigned so much responsibility for the NENYIAC study in earlier years. For the past ten years, the younger Restall had been engaged in private consulting work ranging from basin planning to project design. The commission's response was enthusiastic. Gregg had deliberately moved to fortify NERBC's credibility within the region—yet he had counterbalanced a seasoned public official with a younger professional from the private sector and his own experience in administration, communications, and environmental practice with that of professionals in engineering and water resources.

One other matter of considerable interest came up. The chairman reported new legislation in Congress sponsored by Sen. Birch Bayh (Indiana) designed to encourage regional water resources research. The original Bayh bill would have established a separate institution for each river basin region charged with research responsibilities. Gregg reported correspondence between his office and that of Dr. Bernard E. Berger, representing both the Massachusetts Water Resources Research Center and the New England Council of Water Resources Research Center Directors. As an alternative, he had suggested to Berger a specific liaison for the centers who might be given space and support in the Commission's offices. There was strong encouragement for the idea among Commission members. Water planners and managers needed a mechanism for identifying and responding to urgent research needs. Gregg was authorized to explore the matter further.

1969

NERBC's first meeting of 1969 in Boston found the commission well settled into its regular business. The chairman had heard from Robert D. Brown, the director of the Connecticut Capitol Planning Region, who urged commission opposition to the proposed diversion of the Connecticut River at Northfield. Although any appropriations for the Narragansett Bay Study, now expanded to include all of Southeastern New England (SENE), looked increasingly doubtful, the remainder of the commission's progress reports were encouraging.

Chairman Gregg advised that he was considering taking a "road show" of commission principals and staff to each state during 1969. The response was immediate and positive. Less supportive were the comments on Gregg's explorations with the private utilities of an NERBC project to identify future power sites throughout the region. This could prove to be a highly unpopular topic, the Corps's John Leslie observed.

Accord then vanished entirely when Gregg introduced the subject of a role for the commission in the area of water quality, already the province of Interior's FWPCA, the states, and the region's two interstate compact agencies, NEIWPCC and the Interstate Sanitation Commission (ISC). Enormous sums had just been authorized by the Congress for sewage treatment plant construction, and the Clean Water Act also contained provisions for area-wide waste water treatment planning, another federal single-purpose, comprehensive planning program. Gregg advanced a draft charter for an NERBC water quality committee, arguing that the provisions of Title II of the Water Resources Planning Act mandated a Commission role as the coordinating agency for planning within the region. He was asked if every subject area would eventually require a committee. What about those already served by an interstate agency, such as water quality? the ISC's Thomas Glenn questioned. Speaking for NEIWPCC, Alfred Peloquin wondered whether NERBC intended to duplicate his agency's work in establishing water quality classifications, setting standards, and enforcing implementation schedules. What would the new committee do that is not presently being done under individual studies? John Leslie questioned. The extent of disagreement on his suggestion for a water quality committee was such that the chairman saw no point in protracting the discussion further.

The commission then took up two other vexing coordination matters. One was the set of new authorizations for estuarine studies by FWPCA and the Bureau of Sport Fisheries and Wildlife, both Interior agencies. The other involved a nationwide study of wild and scenic rivers, an assignment Congress had given to the Bureau of Outdoor Recreation. Both promised to impact upon New England. Interior's Mark Abelson promised that he would provide full briefings on each of these at a subsequent meeting.

Dr. A. Ralph Thompson, director of the Rhode Island Water Resources Research Center and chairman of the New England council of such centers, then confirmed his group's interest in a research coordination project.

In May NERBC met in Chicopee, Massachusetts to enable members to visit the Connecticut Yankee nuclear plant at Haddam and the Mount Holyoke, Massachusetts unit of the proposed Connecticut River National Recreation Area. In calling for approval of the minutes of the last meeting, chairman Gregg noted that the matter of a water quality committee was still open for discussion. Connecticut's John Curry said that he hoped not.

A full slate of status reports was received from members and committees, and action was taken on some. An FWPCA enforcement conference had recommended that NERBC coordinate the preparation of a comprehensive water quality plan for Boston Harbor, and NERCOM was moving ahead with its proposed Nashua River Basin water quality demonstration program at the request of the governors of Massachusetts and New Hampshire. Resolutions were passed authorizing limited involvement in both these projects. Efforts were to be "relatively passive,"[74] the minutes stated.

Interior's Mark Abelson had prepared full briefings for the commission on his agency's estuary protection study and its examination of wild and scenic rivers. An ad hoc NERBC committee for the latter study would be considered only after WRC had settled the larger dimensions of coordination needs. Chairman Gregg once again expressed his regret at the failure of Congress and the federal agencies to make more use of river basin commissions as coordinating agents for their regions. NERBC also took up the first two projects referred to it under the licensing procedures of FPC. The Power and Environment Committee would be busy in coming months, for the states were keenly interested in all such matters. Vermont's Lemuel Peet brought forward his committee's first try at the project priorities statement required by the Water Resources Planning Act. The commission approved the document for limited distribution only to WRC and the New England governors.

By June, a special meeting of the commission had to be called in Boston. The issue was one of program and funding for FY 1970. A crisis was at hand because Massachusetts, New Hampshire, and New York had failed to appropriate their share of the NERBC budget. Also, Agriculture, Interior, and the Army all reported severe budgetary cutbacks; and the Congress had failed to appropriate to the authorization level for the much-heralded national water pollution control program. Despite these financial problems, the commission was advised that New Hampshire Governor Peterson had requested NERBC help in evaluating the Seabrook nuclear plant proposal, and Maine Governor Curtis wanted NERBC to sponsor a regionwide study of the coastal zone.

After much discussion a full budget of $376,000 was approved on the assumption that state contributions would be forthcoming. Provision was made for the first year of SENE, NERBC coordination of a coastal study, and a full schedule of activities by the Power and Environment Committee. The commission must always try to be responsive to the interests of the governors, commission members felt.

By the October meeting in Boston it was apparent that faith in the states would be rewarded. Special budgets were under consideration in New Hampshire and Massachusetts, and they looked promising. New York could allocate its funds by administrative action. NERBC's first vice-chairman, Austin Wilkins, would be retiring from state service to be replaced by the director of state planning for Maine, Philip Savage. It was time for the states to elect a new vice-chairman and, in so doing, to review their role in the commission.

Vermont's Lemuel Peet emerged as the states' unanimous choice. A former state conservationist for the U.S. Department of Agriculture in

Vermont, he combined federal experience with a deep interest in state programs. Tall, soft-spoken, and low-key, the new vice-chairman had already displayed his commitment and his capabilities in sensitive assignments such as NERBC's project priorities ranking effort. Peet spoke of the outcome of the state caucus held just before the meeting. The vice-chairman would serve a two-year term from now on. In order to provide leadership for the states, he would hold a meeting of state members separately before each commission session. To expedite a stronger state role, staff assistance would be needed. Chairman Gregg was delighted to make the requested assignment.

The commission then began a program innovation. At each meeting, a special seminar would be held on a topic of widespread interest throughout the region. These would be designed for commission benefit but would also enable outsiders to learn more about its program and activities. The first was on coastal planning. Presentations were made on the work of the Stratton Commission and the new Marine Council, and on FWPCA's national estuarine pollution study and the related Fish and Wildlife Service study. Maine planning director Philip Savage advised the commission of his state's new coastal development planning project made possible by grants from HUD and NERCOM. Chairman Gregg reported a proposal by Sen. Abraham Ribicoff (Connecticut) for a special study commission for Long Island Sound. The senator had been advised that a preferable route might be a comprehensive study under NERBC's auspices.

In other business, a resolution was adopted urging WRC to add the Atomic Energy Commission to membership in NERBC. A prodigal son returned to describe a new private organization just formed to coordinate environmental interests and resolve conflicts within the New England region. Dr. Charles H. W. Foster, former Massachusetts official and activist within the old NRC, was serving as chairman of the board of the New England Natural Resources Center. NERBC's Power and Environment Committee reported a measure of progress in bringing together the NERCOM and NERBC programs relating to energy. The latter agency would consider assigning a portion of its contract funds to NERBC, but the bulk would be spent on a special study of the region's electric power needs to be made by H. Zinder Associates in Boston. In the meantime, NERBC's own study of siting laws and procedures was well under way.

At the December meeting in Boston chairman Gregg welcomed Capt. Roger J. Dahlby of the Coast Guard as the new Department of Transportation alternate. The river basin commission chairmen had met recently with the Water Resources Council in Washington and had encountered "a gratifying degree of interest"[75] on the part of the council members. An appropriation of $200,000 for SENE was included in the Public Works Appropriations Act of 1970.

Vice-chairman Peet reported the results of the states' morning caucus. State members still felt a lack of real involvement in the commission's program, he said. They should be encouraged to prepare water resource plans as elements of the region's planning program. NERBC should provide assistance for these projects and for state participation in its own studies. The states' priorities were reflected still further in a request by the New England governors for an examination of the use of persistent pesticides within the region, a topic that seemed to fall outside of NERBC's technical competence.

The commission then received progress reports on the NAR and NEWS studies. Chairman Gregg again expressed his concern for the narrowness of the NEWS mission. He felt that the public should be informed well in advance of possible engineering solutions. The planning should include an examination of alternatives and effects in the context of balanced, multi-purpose planning. His views were criticized sharply by the members from New York, Connecticut, and the Army. NEWS was a technical study and should go forward without interference from the commission, they observed.

The chairman had the final word in the state planning seminar that followed. Massachusetts and Ohio described their water resources planning programs. There followed progress reports on their use of Title III planning funds from all the states, most of which expected to be engaged in special projects rather than comprehensive water resources planning. Chairman Gregg observed that the key to joint planning was aggressive leadership from the states.

The final commission business for the year related to the Connecticut River Basin. At the conclusion of the Corps's comprehensive study, Jean Hennessey of Hanover, New Hampshire, had been asked to chair a special ad hoc committee to explore a Connecticut River Basin program under NERBC auspices to be supported by supplemental contributions from the states and the federal government. The Hennessey committee's recommendations were accepted with enthusiasm. Connecticut's William S. Wise expressed his delight with the event; he described the program as "almost an ideal approach."[76] NERBC would have its first regional office in a river basin where conflict was almost certain to emerge, for Massachusetts had legislation pending that would set up its own study commission.

1970

At the thirteenth meeting of the commission, held in March in Bedford, New Hampshire, Gov. Walter R. Peterson delivered the key-note address, the thrust of his remarks pertaining to growth, which he characterized as New Hampshire's major problem. This was espe-cially timely because Sen. Henry Jackson (Washington) had introduc-ed legislation broadening the responsibilities of river basin commis-sions to include land as well as water, and hearings were underway in Washington on proposed coastal zone legislation. The governor's remarks were a fitting prelude to a special evening session devoted to strengthening state programs.

Turning to its own business, NERBC learned of widespread reorga-nization activity within the states from vice-chairman Peet, and prog-ress reports were received on the NAR and NEWS studies, and the ac-tivities of the commission's own flood plain management committee. SENE was underway with a $200,000 appropriation. Chairman Gregg also reported that a new NERBC pesticide committee had been estab-lished, as requested by the New England governors, under the chair-manship of Massachusetts commissioner of natural resources Arthur Brownell.

The commission was advised that the Connecticut River Basin hear-ings had elicited rather poor attendance. Citizen interest had appar-ently waned over the six-year study period, and it was felt that educa-tion programs would be needed if the study's major elements were to be carried forward. Gregg reported that all subsequent Type 2 studies would be reviewed by the commission and WRC before being allowed to proceed.

At the Power/Environment Committee's seminar entitled *Resolving Power/Environment Conflicts,* Northeast Utilities chief executive officer Lelan F. Sillin, Jr. described his company's siting procedures. He was followed by Charles H. W. Foster of the New England Natural Re-sources Center who described Project CEVAL, the experimental effort in western Connecticut and Massachusetts to evaluate openly North-east Utilities's pumped storage facility with the help of a broadbased citizen committee. Judging from the seventy-five people present, power and environment would be a topic of considerable interest within the region for some time to come.

At the June meeting of the commission in Boston, the chairman re-ported issuance of Presidential Executive Order 11528, which, at long last, extended the commission's jurisdiction to Long Island Sound and added the Atomic Energy Commission as a new member of NERBC. He noted that full appropriations had been received from all of the states. After some discussion, the members adopted budget resolutions call-ing for a 10 percent increase in state support to cover inflation and a three-month carry-over reserve to allow for appropriations delays. Gregg was also authorized to bring before the governors the need for a special $120,000 per year follow-up program for the Corps's Connecticut River comprehensive, which would include the citizens review committee requested by Rep. Silvio Conte (Massachusetts).

On power/environment matters, a number of events had transpired since the March meeting. Gov. Francis W. Sargent (Massachusetts) had officially requested NERBC assistance in evaluating the proposed Northeast Utilities' pumped storage site in western Massachusetts. NERBC's Power/Environment Committee had also completed work on a two-year, $500,000 study proposal that would be sent to WRC for review and approval. Less promising was the attitude of the House Appropriations Committee, which had questioned the lateral transfer of funds from NERCOM to NERBC for such projects as power plant siting and evaluation. The committee felt that its own prerogatives were being curtailed by such administrative transfers.

But the bulk of the commission meeting was devoted to the topic of planning. Vice-chairman Peet spoke of the states' strong interest in coordinated, joint, state-federal planning that would serve as the basis for both state and federal actions and thereby provide the states with "more nearly equal footing." John B. Roose, assistant director of WRC, advised that the old numerical listing of federal studies had been replaced with the letter designations A-D. The Type B or comprehensive studies would require less data gathering than the old Type 2 studies and, hopefully, pay more attention to management needs. There was still no progress in achieving the objective of centralized funding of all studies through WRC, he reported. This triggered extensive discussion of the commission's own responsibilities for preparing a CCJP for the region. State members were adamant that the federal government must accept the states as the primary region for planning and decision making. If not, theirs would be purely a passive role, a situation that Connecticut's William Wise described as unsatisfactory during NRC. New England's CCJP would consist of three elements, the NERBC finally concluded: 1) a framework plan derived from the Corps's NAR study; 2) a series of plans developed by the individual states with commission staff; and 3) a number of management plans for subregions such as the Connecticut River Basin, SENE, and Long Island Sound as Type B funds became available. Formulation teams would be constituted for each study, and federal funds would be requested from WRC to cover NERBC expenses.

At the September meeting of the commission in Brattleboro, Vermont, NERBC was advised of the application of its new ground rules to the SENE study, which would be the first of the so-called Type B comprehensives under WRC's new policy guidelines.

Chairman Gregg could also report a good session with the New England governors. They had approved the commission's two-year budget and also supported a special Connecticut Basin program to be funded by special appropriations of $35,000 to be shared equally by the four basin states and then matched with federal funds. In their capacity as NERCOM commissioners, the governors had endorsed the establishment of a special New England energy policy staff within NERCOM as recommended originally by NERBC's earlier laws and procedures study.

But matters were continuing to simmer in the Connecticut Valley. The Corps's draft plan for the basin had recommended a number of controversial flood control dams, notably the Gaysville and Victory dams in Vermont and the Meadows Dam on the Deerfield River in Massachusetts. The last had caught Congressman Conte's attention and had triggered his recommendation for an independent citizen review of the entire Corps plan, particularly during the pending ninety-day review period. Four hydroelectric power licenses in the valley were scheduled for review by the FPC. State legislation in Massachusetts authorizing a water supply diversion from the Connecticut by way of the Northfield Mountain pumped storage facility had been stalled by rising citizen protests from Massachusetts and Connecticut. This had prompted Massachusetts House Speaker David Bartley to sponsor a study commission to explore ways of building Connecticut Valley interests more securely into basin planning.

At a special evening session, NERBC members were apprised of Vermont's new approaches to land use. They also heard that a special study conducted by Anderson-Nichols had revealed that flood plain management regulations developed by the University of Wisconsin could be adapted to New England at a cost of $100,000 per state.

For the December meeting in Boston, the year-long discussions about planning concluded with a special seminar with NRBC and outside panelists from the Conservation Foundation and the Coastal States Organization, the primary lobbying group for the pending federal coastal zone legislation. Nearly eighty New England leaders attended.

At the business meeting, chairman Gregg reported the activation of two new federal agencies: the Environmental Protection Agency (EPA) and the National Oceanic and Atmospheric Administration (NOAA). Each would have an impact on the commission's future programs. He advised NERBC members that the special pesticide committee had completed its work and transmitted a model act for pesticide regulation to the New England governors. On the subject of planning studies, there would be no funds for the Long Island Sound Level B until the FY 1971 supplemental budget was approved by the Congress. However, Sen. Abraham Ribicoff (Connecticut) was taking a personal interest in the matter.

Dr. Ervin H. Zube, director of the University of Massachusetts's Institute for Man and Environment, then described his concept of an environmental quality plan as consisting of elements of environmental diversity, environmental accessibility, and continued resource quality. These principles would be applied within the SENE study, the chairman advised. Dr. Rudolph W. Hardy, the newly-appointed research coordinator for the New England Council of Water Resources Research Center Directors, gave a brief presentation of the steps he would take to achieve research coordination. He expressed appreciation for NERBC's action authorizing the establishment of a research priorities committee to work with his council.

The remainder of the business session was devoted to a report from Dr. Bernard E. Berger, director of the Massachusetts Water Resources Center, who was serving as chairman of the Connecticut River Basin Citizens Review Committee. He advised that six subcommittees had been formed, the work was going well, and he planned to submit the committee's recommendations to a reconvened meeting of the Corps's Connecticut River Coordinating Committee some time after the first of the year.

1971

On February 9 NERBC convened a special meeting in Amherst, Massachusetts to receive the report of Berger's committee.[77] Public attendance was slim, and only Vermont and Massachusetts sent state representatives. Dr. Berger stated that ten elements of the coordinating committee report had been examined by his six subcommittees and agreement had been reached on all but one. A minority report on flood control had been submitted expressing a stronger position against upstream dams. All in all, Berger observed, this was a worthwhile effort and a good device. Outside comment was received expressing concern about further diversions from the Connecticut River. The Berger Committee's inclusion of this possibility only if there existed "truly surplus waters" and under "controlled" circumstances seemed generally acceptable. Connecticut's Clyde Fisher urged prompt attention to the flood protection needs of urban areas. Springfield's Victor Gagnon, a veteran Connecticut Valley warrior, spoke of the need for a continuing advisory group and possibly a basinwide action agency. Gregg reminded the group that the Connecticut River Comprehensive Report was not a project-authorizing document in itself. WRC would undoubtedly focus on its early action plan provisions for any subsequent implementation.

NERBC resumed its regular schedule in March in New Haven, Connecticut, with an open session on the commission's Long Island Sound Study (LISS). As study manager David Burack described the program, LISS would be a three-year planning effort involving a team of agency professionals under the guidance of a state-federal coordinating committee. A parallel nongovernmental structure would be established to advise on citizen concerns and enable community participation. The plan would be prepared from a series of vertical or single-purpose studies conducted by special agencies. NERBC would coordinate the entire venture, receiving $22,000 of the initial $100,000 in planning funds. The remainder would be appropriated directly to participating federal agencies. LISS would be meaningful because, as Rep. Lester Wolff (New York) assured the approximately one-hundred persons present, there would be a moratorium on federally supported projects within the sound until the study was completed.

At the commission's business sessions, chairman Gregg introduced representatives of the newly authorized EPA, Bartlett Hague and Patrick Nixon. He reported pending legislation to provide the advance consent of Congress to interstate environmental compacts. Gregg also advised of an initiative by the New England governors to strengthen the existing structure of regional cooperation in the environmental and natural resources fields. A special committee consisting of himself, executive director Chapman Stockford of the New England Governors Conference, and Charles H. W. Foster of the New England Natural Resources Center would develop recommendations for the governors. He also reported a forthcoming conference, jointly sponsored by WRC and the Interstate Conference on Water Problems, on strengthening water planning—a sequel to the session held a year ago under similar auspices reviewing the first five years of the Water Resources Planning Act.

Committee reports indicated that the NAR study was winding up and that NEWS consulting reports had been received describing the water supply alternatives for New York, Washington, Boston, and eastern Massachusetts. The SENE Coordinating Committee had settled on the Ipswich River Basin in Massachusetts as its pilot planning area. NERBC's analysis of the Connecticut River Comprehensive Report was scheduled for submission to WRC by May 1, 1971.

The June meeting of the commission in Boston revealed several changes. It was announced that Robert D. Brown, former regional planning agency director in Hartford, Connecticut, would join NERBC as staff director effective September 1. State members caucused to elect Arthur W. Brownell of Massachusetts as the new vice-chairman. But much of the session was devoted to budgetary considerations for FY 1973. Planning budgets for SENE ($1,318,000) and LISS ($3,005,000) were approved—also a $300,000 request for a bulk power supply study which was subject to further review by WRC. It was likely that NERCOM would assume a measure of responsibility for the facility siting phases if proper coordination could be arranged. An FY 1973 NERBC operating budget of $431,000 was also approved.

State members reiterated their concern over lack of participation in the commission's planning studies, for none of the funds available from WRC could be directed to state participants. This left them unable to bring their urgent problems before NERBC. Following extensive discussion, a new series of state guide plans was recommended for Maine, New Hampshire, and Vermont to parallel the attention being given to southern New England problems by the SENE and LISS studies. These would be funded from the commission's operating budget and consist of one NERBC staff member working directly with a state professional supported by Title III water planning funds. Each planning project would be activated upon specific request by the governor of that state. It was further recommended that the NERBC seek WRC funds for a joint study of the remainder of New England's coastline (New Hampshire and Maine).

Among other matters of interest, the Corps of Engineers reported initiation of a waste water study for the Merrimack River, a pilot program under the auspices of NEWS. The current assistant secretary of the Army with responsibility for the Corps had been much impressed by the waste water renovation experience at Muskegon, Michigan. The HUD representative also spoke of his agency's interest in coordinating the HUD 701 planning program with the comprehensive studies being undertaken by NERBC. He suggested that a handbook for local officials might be developed jointly by the two agencies using a consultant selected by both agencies but reporting directly to NERBC.

Commission members traveled to Kennebunkport, Maine, for the October meeting, where an informal conference took place on the northern states' guide plan program. State members estimated that each plan might require two years' time and approximately $150,000 in funds. The need for flexibility was stressed to accommodate the particular problems and approaches in each state. Active involvement of the substate regional planning agencies was recommended. Commission members also heard from the four states undergoing reorganization of natural resources and environmental programs: Connecticut, Maine, Massachusetts, and Vermont.

In other business, chairman Gregg reported that Sen. Abraham Ribicoff (Connecticut) had scheduled oversight hearings earlier in the month on LISS and seemed generally satisfied with progress. Massachusetts Gov. Francis W. Sargent had recommended the inclusion of all of Massachusetts and Vermont within the NERBC's jurisdiction, an item that the commission approved for subsequent action by WRC. Gregg also advised that two separate citizen bodies had been established for the successor Connecticut Basin Program: a Citizens Advisory Board and a Science/Research Advisory Committee. A private organization of concerned scientists, the Connecticut River Ecology Action Corporation (CREAC), had also been formed. Commission members learned of Rhode Island's new Coastal Resources Management Council, established to strengthen control over developments in Narragansett Bay and other coastal regions, and of a HUD/U.S. Geological Survey national urban pilot program, an effort to provide local decision makers with relevant land use information. New England's initial project in the Hartford-Springfield area had proved so successful that it had been extended northward to Brattleboro, Vermont and southward to Long Island Sound.

Chairman Gregg triggered extensive discussion by sharing with the commission a memorandum from Massachusetts director of mineral resources Robert Blumberg to his commissioner, Arthur W. Brownell, arguing that New England needed to work together to protect its interests should development of the Outer Continental Shelf's oil and gas resources take place. He enclosed a recent press release from the Department of the Interior announcing its intent to proceed with leasing on Georges Bank. NERBC authorized the chairman to contact Interior Secretary Rogers Morton and the New England governors to explore ways by which the commission could become involved.

1972

As the commission convened in Boston for its February meeting, a number of energy developments could be reported by chairman Gregg. The nuclear plant at Seabrook, New Hampshire, had been reactivated as a project by Public Service of New Hampshire, but its future was no more certain than before. The New England Energy Policy Study was underway under NERCOM auspices and the directorship of an old friend of NERBC, Paul Shore of FPC. More disappointing was the Office of Management and Budget (OMB) decision to disallow the proposed bulk power study, a cooperative venture between NERCOM and NERBC, albeit "without prejudice."[78] But the two regional agencies had entered into a joint contract with the University of Rhode Island to examine the relationship between water and economic development in the SENE area.

Gregg could report the activation of a thirty-member citizens advisory committee for LISS. He urged that the federal agencies consider subcontracting some of their study funds to the states of New York and Connecticut to enable increased state participation. As for the Connecticut Basin study, it was in trouble, Gregg reported. No federal funds were included by OMB in the President's FY 1973 budget, and the states of New Hampshire and Vermont were in default on their commitments.

Among agency and staff project activities, the Nashua River demonstration and Boston Harbor projects were reported as continuing. The costs of harbor clean-up were now estimated at $2 billion. The Corps of Engineers had completed its Merrimack waste water study, but the next step, a survey scope study, would not take place because of New Hampshire's concern over a possible diversion of water to Boston. Massachusetts hoped to persuade the Corps to expand the study to include the water supply and waste disposal needs of the entire metropolitan area.

Under new business, inclusion of the entire Lake Champlain drainage area in the commission's jurisdiction was discussed and deferred in the light of differences in attitude between New York and Vermont, and concern was expressed over WRC's proposed principles and standards. In response to discussions at the October meeting, the chairman reported that an ad hoc steering committee of the two regional agencies and the states was prepared to recommend a joint NERCOM/NERBC study of the economic and environmental implications of Outer Continental Shelf development.

The May meeting of the commission was held at Chatham, Massachusetts with MIT Professor Jay Forrester of *Limits to Growth* fame as the featured speaker. Chairman Gregg was scheduled to leave right after the meeting to join his fellow river basin commission chairmen in Washington for a meeting with the Council of Representatives of WRC. He reported that the chairmen had arranged a special meeting with OMB officials to discuss budget and coordination matters. They were informed that OMB had asked three of the federal regional councils (all outside New England) to undertake trial projects leading to improved federal coordination. Gregg also reported that Robert Restall would be leaving the commission staff for an assignment with FPC in Washington. His place would be taken by a Connecticut water professional, Irving Waitsman, who would be asked to head NERBC's new Technical Services Division.

The Corps of Engineers reported several recent developments. Four pilot areas had been selected nationally (Vermont among them) to explore improved compliance with the National Environmental Policy Act. Further, a deep water ports study had been authorized by Congress to extend from Eastport, Maine to Hampton Roads, Virginia, with four possible sites in New England. Rhode Island's Daniel Varin indicated that the Corps had also been asked to consider possible involvement in the Big and Wood River reservoir projects in order to ensure multipurpose considerations beyond the water supply needs of Providence.

Reports were received of a national study by EPA of eutrophication problems and a parallel study of New England lakes submitted by the Council of Water Center Directors to the Office of Water Research and Technology.

In other matters, NERBC learned that it would be asked to conduct the regional phase of the national water assessment by WRC. It endorsed a joint request by New York and Vermont for a Level B study of Lake Champlain. The commission was advised that all state pledges for the Connecticut Basin Program were now in hand and that the project would be activated in FY 1973. The prospects for federal coastal zone legislation remained excellent in the opinion of most observers.

Chairman Gregg brought up his suggested consolidated coordinating group for the policy aspects of all major studies, pointing out the inconsistencies inherent in the present setup. However, commission members continued to prefer arrangements tailor-made to the individual studies since these ensured less centralized authority. Gregg agreed to maintain the present procedures but indicated that he would work toward better definition of the authority of the coordinating groups and study teams.

At the September meeting in Boston Leonard Crook, executive director of the Great Lakes Basin Commission, was an observer. Chairman Gregg indicated that all river basin commission chairmen and the Council of Representatives of WRC would be meeting in New England in October. He asked Maine representative Philip Savage to report on the establishment of the new Maine Department of Environmental Protection and spoke of his own involvement in the Massachusetts reorganization task force on energy.

New York's Jack Finck expressed concern about the slow progress and excessive inventory activity under LISS. In related action, NERBC disapproved a request for a voting position on the LISS coordinating committee for the citizens advisory committee, but permitted a designated member to participate ex officio. No citizen appointments had been made as yet by the governor of New York, it was reported.

John Leslie of the Corps of Engineers advised the commission of extensive opposition to the idea of a single regional port facility to handle petroleum shipping. The primary obstacle was the intense competition among New England deep water ports. He advised that the NAR and NEWS studies were in their final stages and that the Charles River comprehensive study had been completed. Since the last meeting, the Corps had been assigned nationwide responsibility for the inspection of dams and had set up a new advisory committee on ocean dumping.

In final action, commission members approved the draft study proposals for the northern states guide plans. State members caucused after the session to discuss a revised apportionment formula for their share of NERBC operating expenses and to hold a preliminary session on strategies for natural resources decision making.

By the December meeting of the commission in Boston, the national elections had ensured another four years in office for President Richard Nixon. Chairman Gregg reported that his resignation had been submitted pro forma. As for the commission, its study of regional pumped storage sites had been completed. Of the fourteen sites identified, five were in the category of least environmental impact. LISS had still not advanced to the plan formulation stage, Gregg reported, due to funding and procedural delays on the federal side. An interim report on dredging was available, however, which could be used in court cases. The commission was also advised of difficulties with NERBC's share of the national water assessment because of insubstantial funding by WRC. Gregg had dismissed the commission's water supply committee for lack of assignments but felt that the committee on flood plain management should remain on a standby basis. New York's Jack Finck reported on his state's recent experience with flood plain hydrology studies, which Gregg cited as a possible pilot project for the New England region.

Next on the agenda were various matters relating to marine and coastal resources: completion of the Penobscot Bay study, a pilot program for the entire Maine coast; the new ocean dumping law, which contained provisions for research and the designation of marine sanctuaries; and the possibility of a new NERBC coordinating committee for dredging and ocean dumping. Two candidate New England petroleum deep water ports had emerged from the Robert R. Nathan Associates national study—Portland and Boston harbors—which the Corps of Engineers was encouraged to pursue. Gregg also reported NERCOM's interest in establishing a New England marine resources council to encourage the economic development and utilization of marine resources.

NERBC then convened a second session to discuss a draft staff report on strategies for natural resources decision making, which was scheduled to be submitted to the New England governors on December 15. The chairman reminded the commission of the governors' interest in achieving better use and coordination of federal planning programs. He felt that NERBC should suggest to the governors that new programs be put together within the states and the region in such a way that they produce for each state an integrated program for managing natural resources. In response to questions from New York and Maine, Gregg stated that he visualized the NERBC role as that of coordination and integration at interstate and federal levels, not as intervenor between state and federal program agencies. EPA Regional Administrator John A. S. McGlennon, the current chairman of the Federal Regional Council, predicted that it would soon be moving into the natural resources area, particularly in the administration of grants. Gregg saw no need for another natural resources program if the coordinating functions of the existing regional agencies were themselves properly coordinated, citing the operating understandings for natural resources decision making approved for distribution to the governors with the clarifications requested by Maine and New York. The final document contained the operative language most sought by chairman Gregg: The commission would act hereafter as the "principal agency for coordination within the region of plans for management of water and land resources."

1973

The March meeting of NERBC was held in Waterbury, Vermont, an inauspicious occasion for more reasons than the inclement weather. Commission members had learned of sharp cuts made in many programs by the incoming Nixon administration. The prevailing sentiment was: What can NERBC do to help? Faced with reductions of as much as one-third in Bureau of Outdoor Recreation and Title 701 funds, some state members felt that their programs would be set back by ten years. Despite congressional passage of the coastal zone management act, there had been no appropriation as yet for its implementation. There was a faint ray of hope, however. According to deputy associate director Richard M. Vannoy of WRC, Section 209 of the Federal Water Pollution Control Act amendments had settled a matter of long-standing concern to the commission. When and if moneys are appropriated for new Level B studies, the program would be centrally funded through WRC.

After receiving reports on all major programs, the commission turned its attention to the still vexing program of planning coordination. Massachusetts Assistant Secretary of Administration and Finance Robert H. Marden, the current president of the National Association of State Planning Agencies, spoke of his state's struggle to bring together water quality planning with other forms of regional planning. The situation will grow even worse when coastal zone planning funds come available, he noted. The commission offered its full support for the adoption of uniform guidelines by WRC.

After approving budget levels for FY 1974 and FY 1975, the commission adjourned to a session on Vermont's environmental programs with special attention to the planning and regulatory activities under its Act 250.

The commission traveled to Bedford, New Hampshire for its June meeting. A special guest was David Kinnersley, an observer from the British Association of River Authorities. Chairman Gregg reported that Massachusetts's Arthur W. Brownell had agreed to serve as vice-chairman for a second two-year term. He advised of a recent special meeting of federal members that had discussed, among other topics, the possibility of state contracting for studies. He noted that the New England governors had decided to broaden their conference to include participation by the eastern Canadian premiers, an event of potential significance in the energy and water resources fields.

Gregg then advised the commission of the President's decision to zero-fund NERCOM and all similar Title V commissions. This action was likely to be overturned by Congress, but the governors were exploring a contingency plan to form a private organization, NERCOM Inc., to continue its work in energy and transportation. He had also learned that the Federal Regional Council had included a natural resources element in its FY 1974 work plan. Faced with these developments, the special committee on regional cooperation in environmental and natural resources fields had submitted a supplemental report to the May 21 meeting of the New England governors containing four recommendations: 1) that a strong statement supportive of regionalism be adopted; 2) that any residual NERCOM funds be used to strengthen and expand the staff capabilities of NEGC in the environmental area; 3) that the federal government be urged to utilize NERBC as a primary coordinating environmental agency for New England; and 4) that the governors and state members of NERBC strengthen state leadership in the commission so that it could serve as the focal point for environmental matters in the region.[79] The last recommendation was deemed especially important. It could be achieved by having a natural resources/environmental professional on the staff of the Governors Conference itself, and by having that individual serve as full-time vice-chairman or, at least, full-time staff to the elected vice-chairman.

In general discussion, state members recalled the questions they had raised at their March caucus as to what kinds of regional mechanisms would be needed to deal with the environmental and natural resources issues of the next decade. The special meeting of federal members had considered a related topic: possible polarization of state vs. federal viewpoints if the states should opt for a mechanism to which the federal agencies were not a party. There was general agreement on the need for a study of regional institutional arrangements, but no clear consensus as to how such a study might be accomplished.

Turning to program matters, the SENE, LISS, and Connecticut River Supplemental Study (CRSS) projects were described as functioning well but behind schedule. NERBC's Boston Harbor Committee had been disbanded in favor of a new mechanism established by Massachusetts to develop plans for waste water management in the entire metropolitan region. Responsibility for the Nashua River demonstration project had devolved to NEIWPCC. Level B study funding would be requested by the administration, but of the three starts included for FY 1974, none was in New England. NERBC's three possible new Level Bs—Maine–New Hampshire Coastal, Lake Champlain, and the Housatonic–Thames—could be considered for the FY 1975 budget. He said, however, a potential complication had arisen with the Corps of Engineers' new authorization for urban studies projects. There was congressional interest in such a study for the Housatonic.

With respect to internal matters, chairman Gregg reported that he had followed through on his objective of streamlining study procedures. The staff director had been assigned the job of coordinating the three comprehensive studies. He asked for NERBC approval to allocate funds and enter into contractual arrangements directly with the states for specified study tasks. Agriculture's Charles Dingle objected to any diversion of funds appropriated directly to a federal agency, but the other federal agencies agreed to the concept in principle if it could be accomplished legally.

The chairman also advised that Vermont and Maine had contributed less than the full amounts due the commission. The only recourse was the reduce NERBC's operating budget. The commission then adjourned to a special session devoted to the Council on Environmental Quality's new responsibilities for studying the environmental impact of offshore oil development, and to a briefing on the status of the pending national land use policy legislation.

By the September meeting in Boston the shortfall of state appropriations had grown to include New York. However, NERCOM funding for FY 1974 had been restored, as predicted, by the Congress. It was also learned that the President would seek appropriations for the coastal zone management program, a matter of considerable interest to six of the seven NERBC state members.

At long last, LISS was in the first stage of plan formulation. Progress was reported on SENE and the Connecticut River Supplemental Study. NERCOM had developed an energy program that included the long-sought bulk power siting study. Of particular interest to commission members was NERBC's proposed study of flood plain management efforts in New England, which would be submitted to WRC for funding in FY 1976. It included the application of flood plain management techniques to a number of pilot communities, thus raising general concern over intrusion of NERBC into the established prerogatives of state and federal agencies. The commission was also advised of its role in the national assessment required of WRC, an activity that could help NERBC develop its statutory CCJP.

Assistant director Gary D. Cobb of WRC then briefed NERBC members on the new principles and standards for the planning of water and related resources. Every federal program must hereinafter address two national objectives: economic development and environmental quality. Two additional areas may be considered: regional economic development and social well being. Thus, every federal study (except those "grandfathered" by the head of the affected agency) must include at least two alternate plans. He added that EPA and HUD did not presently subscribe to these policies and that they, naturally, did not apply to state programs.

In final matters, the commission was advised that the earlier matter of a study of regional institutions was still very active. There were indications that the Council of State Governments might be able to provide assistance. Also, a major regional conference was being planned to discuss and determine the major regional development issues facing New England.

At the December meeting of the commission in Boston a special guest was Warren Fairchild, the new director of WRC, who spoke of his efforts to streamline Level B planning and to tackle the problem of cost sharing for water projects. He advised that, in the future, individual proposals would proceed through two preliminary stages: Proposals To Study and Proposals Of Study, each of which would be funded by WRC. In response, the NERBC staff had proposed a new series of "sketch plans," which would identify major problems and opportunities and make some general policy recommendations for a particular basin at one-third of the time and cost of the conventional Level B study. The response from commission members was generally favorable.

Turning to the commission's own business, the chairman reported a recent session chaired by NERBC at Connecticut's request to consider the matter of future diversions from the Connecticut River. He reported seven public flood management forums held in the Connecticut Valley to explore what would have happened if Hurricane Agnes had reached northward into the basin. Gregg spoke positively of the northern guide plans, feeling that there was a greater return from money invested than from the coordinated agency studies.

Discussions then resumed on the coordination of various natural resources planning programs. EPA's Walter Newman described the current situation with respect to his agency's Section 208 planning, adding that EPA was not optimistic about future funding. Joseph Ignazio then described the Corps's new urban studies program, reporting that, in New England, the Merrimack and Boston waste water studies were now being carried under the program and others were planned for the Housatonic (Massachusetts and Connecticut) and for Pawcatuck–Narragansett Bay (Massachusetts and Rhode Island) if state participation could be assured. Left unanswered was the question of whether a Corps urban study could qualify as the Section 208 study required by EPA.

In other matters, the NERBC staff displayed a situation map of New England pinpointing hundreds of regional natural resource issues. One potential area of dispute included problems of flooding along the Richelieu River, the outlet for Lake Champlain on the Canadian side of the border. The suggestion of artificial lake level management for Lake Champlain was a matter of great concern to New York and Vermont, and a current responsibility of the International Joint Commission, a bilateral agency involving Canada and the United States. Another potential problem was the matter of dredged spoil disposal from coastal beaches such as Long Island Sound. Lack of sites was hampering the Corps of Engineers in its routine harbor and channel maintenance activities.

NERBC members then returned to the matter of regional institutions. They reviewed the chairman's recent correspondence with William F. Powers, the chairman of the Federal Regional Council, in which Gregg suggested a system of mutual observers rather than a formal FRC natural resources committee. It was reported that the state members would await the outcome of the regional natural resources conference, now scheduled for February, before developing recommendations for an NERBC-sponsored study of institutional options.

The commission heard from Daniel Varin, former Rhode Island planning director, now on leave to head the Council of State Government's national land use project. Varin had drafted a thoughtful paper analyzing the needs and opportunities for regional institutional arrangements, which had been distributed to NERBC members. In the discussion that followed, NERBC state members made it clear that regional institutions should not interfere with state-local relationships. They felt that the initiative for federal-state coordination and cooperation must lie with the states rather than the regional agency. Its primary function should be to provide effective access for individual states to federal agencies and programs.

Commission members then grappled with program activities for FY 1976. The staff had identified three alternative study directions: 1) new concept Level Bs focused on critical issues rather than geographic regions; 2) basin sketch plans limited to major resource issues; and 3) a regional sketch plan for New England as a whole. Alternatives 2 and 3 would be solely an NERBC responsibility; alternative 1 would require substantially more time and money and state matching funds. Consensus could not be reached, and a special meeting was arranged to follow the March meeting devoted solely to program and budgetary matters.

1974

As the commission assembled for its March meeting in Boston, it learned that much of its optimism was unfounded. The President's FY 1975 budget contained no money for new Level B starts. Closer to home, there were still funding problems in northern states like Maine. Another fiscal problem was confronting the region, Maurice D. Arnold, Bureau of Outdoor Recreation regional director, advised the group. Over half of the unobligated funds nationwide under the Land and Water Conservation Program were in New England states. Chairman Gregg offered to convene a meeting of state administrators to discuss the problem.

In other matters, all of the NERBC studies were reported as progressing satisfactorily. Sen. Abraham Ribicoff (Connecticut) continued to take a personal interest in LISS by holding another oversight hearing in Washington March 18.

A special panel on coastal zone cooperation was then convened. Dr. Perry R. Hagenstein, executive director of the New England Natural Resources Center, urged that NERBC be given regulatory responsibility over coastal developments of regional significance. Maine state planning director Philip Savage flatly disagreed. The state members had caucused to discuss coastal zone matters and favored only a liaison role for NERBC. Nevertheless, a special meeting of state members would be held on April 17 to consider ways of improving regional cooperation in coastal zone matters. Despite staff urging, a follow-on program for the Long Island Sound area was rejected. With New York still uncommitted, the Level B proposal for Lake Champlain had to be deferred. A Level B study for the Housatonic-Thames basins was postponed pending the outcome of the Corps's urban studies in the Housatonic effort. Despite New Hampshire's reservations about possible future diversions, a Level B study for the Merrimack was approved for study design. Among program alternatives, commission members favored the basin sketch plan approach using NERBC operating funds exclusively.

Pursuant to the meeting of state coastal zone officials the day before, it was agreed that NERBC staff would be assigned part-time to help organize and operate an informal coastal zone task force. The action was spurred by a report that NOAA's Office of Coastal Environment would expect multistate cooperation as a prerequisite for grants.

State members also reported agreement on a revised apportionment formula for FY 1976. A new base of $18,590 would be established for each state. The remainder of NERBC operating costs would be apportioned 10 percent by area and 90 percent by population. Because of its limited involvement in the commission region, New York would be charged the base rate only.

Having voted to cancel the regular June meeting, the commission next met in Woodstock, Vermont, in September. It received progress reports on all continuing studies and accepted guide plans for the state of Vermont and the Androscoggin River Basin. Chairman Gregg informed the group that, despite his best efforts, the Federal Regional Council had proceeded to establish a task force on land use and natural resource planning. NERBC staff were participating in it, and at least it had been identified as a joint project of the two regional agencies. He and HUD representative Sheldon Gilbert agreed that there was virtually no likelihood of consolidation of federal planning grant programs. Federal agency jockeying was just too extensive at the Washington level. The best chances of effective coordination lay within the regions. Any optimism in this regard was reduced by reports from EPA, which was having difficulty getting its Section 208 program off the ground, and the Corps of Engineers, which could report no resolution of the relationship between its urban studies program and other planning efforts. In fact, no agreement had as yet been reached with Massachusetts on the proposed Pawcatuck–Narragansett Bay study.

Rhode Island's Daniel Varin then introduced Ralph Widner, president of the Academy for Contemporary Problems, which had agreed to undertake the study of alternative regional institutional arrangements discussed at earlier sessions. The academy would do so under the direction of the state members but with funds provided from the NERBC's operating reserves. Widner agreed to have a final report available to the commission by March of 1975.

The commission also heard from Dr. Charles H. W. Foster, secretary of the Massachusetts Executive Office of Environmental Affairs, who urged a special NERBC program on Outer Continental Shelf development. He felt that the NERBC was the ideal mechanism for drawing together the resources of the federal and state agencies on this issue. The first step should be an options document to help guide pending state policy decisions, but ultimately, the commission should consider the entire range of environmental issues. Commission members responded favorably to the proposal and reconstituted the Power/Environment Committee as the Energy/Natural Resources Committee with Foster serving as temporary chairman.

At the December meeting in Boston Foster returned to recommend an expanded involvement—a new Outer Continental Shelf (OCS) Task Force under a joint federal-state chairmanship. It would establish working groups and require up to two man-years of staff time from NERBC. He felt that the issue was of such significance to New England that a commitment of this magnitude was warranted. State members favoring OCS development expressed concern that their own policy positions might be compromised. Chairman Gregg pointed out that an OCS Task Force would only complement the previously authorized coastal zone task force by providing an offshore orientation. It would deal factually with development proposals and operate at all times within a policy framework established by elected officials. With these reassurances, the proposal was adopted and the chairman was instructed to transmit this information to the New England governors.

In other business, the commisson moved officially to endorse a ⸹750,000 study of Lake Champlain by the International Joint Commission as a prerequisite to any flood regulation program. It also heard that the President was likely to submit national land use legislation to the Congress if the jurisdictional rivalry between Interior, Agriculture, and HUD could be constrained.

Kenneth Rainey and Michael Brewer of the Academy for Contemporary Problems then outlined their suggested approach to the region institutions study commissioned by the NERBC, and a variety of responses was forthcoming. Maine's Philip Savage mentioned the lack of an umbrella agency to bring technical people together with policy makers. By way of illustration. Varin observed that NERCOM could reach policy makers but lacked the relationship with federal agencies enjoyed by the NERBC. Savage also cited the need for a comprehensive view encompassing social as well as environmental and economic issues. HUD's Sheldon Gilbert observed that a variety of regional institutions might, in fact, be the most appropriate response to a variety of regional needs.

The regular meeting of the NERBC then adjourned to enable participation in two special sessions: a panel discussion of water-recreation relationships, and a session on recent developments in land use planning and control.

1975

By the time of the March meeting in Boston state elections had produced a new crop of participants in NERBC affairs. The states had elected as the new vice-chairman Arthur A. Ristau of Vermont, who would bring a special dimension to the position, for he was also serving as the NERCOM cochairman's representative in Boston. Massachusetts's new environmental secretary, Evelyn F. Murphy, had succeeded to the chairmanship of NERBC's coastal zone task force. Much of the commission's focus appeared to be shifting seaward as its two major planning studies, SENE and LISS, entered their final report review, and matters relating to offshore oil development dominated the spotlight. NERBC's new OCS Task Force was reported as participating actively in a conference that would bring together skeptical scientists to examine weaknesses in baseline information off the New England coast.

Richard N. Vannoy of WRC could also report a number of developments on the Washington scene. HUD, Commerce, and EPA had been added to the council and HEW deleted from membership. Sen. Frank Church (Idaho) had scheduled oversight hearings on all titles of the Water Resources Planning Act before his Senate Subcommittee on Energy, Research and Water Resources. Vannoy reported new authority for nonstructural alternatives to flood control contained in Section 73 of the Water Resources Development Act of 1974. Section 80 of that same act called upon the Water Resources Council to conduct a study of planning objectives, cost-sharing and discount rates as they related to water projects.

NERBC staff director Robert Brown reported that federally supported planning programs in New England would reach a level of $32 million during FY 1975. Funds were clearly more than adequate if proper coordination was achieved. In that regard, Joseph Ignazio of the Corps of Engineers advised that an agreement had finally been reached with EPA on urban study relationships. EPA's Bartlett Hague expressed a similar interest in developing coordinated relationships with coastal zone and Title 701 comprehensive planning programs. HUD's Sheldon Gilbert reported that his agency and EPA were currently negotiating a cross-agreement on such matters. He advised that HUD had already agreed to recognize land-use plans developed under the coastal zone program in lieu of its own Title 701 requirement.

Despite these and other encouraging developments, the state members of NERBC displayed little interest in nominating further Level B studies. The Level C or project formulation studies were becoming the dominant activity at federal and state levels. Chairman Gregg suggested that the time might have come to concentrate NERBC's attention on improving state natural resource planning. He offered to hold a series of meetings in the state capitals to explore such a program.

The sensitive antennae of the Corps of Engineers had spotted still another set of opportunities. Ignazio reported receiving several requests from communities for studies of hydro power potential. This could become a new and possible major issue for the future, he observed.

Regional director Maurice D. Arnold of the Bureau of Outdoor Recreation returned to his earlier concern about lagging recreation programs in New England. The commission encouraged him to explore the suggestion of a new standing committee on recreation planning with the state liaison officers.

Because of impending budget matters, a special session was held on May 1 to deal with the commission's program and budget for FY 1976 and FY 1977. A brief series of reports on ongoing programs preceded the budget deliberations.

Rhode Island's Daniel Varin advised that a second draft report on regional institutions had been received from the Academy for Contemporary Problems. Since it was wide of the mark, the state members had authorized vice-chairman Ristau to negotiate with a third party to draft an acceptable paper for the governors. Chairman Gregg reported that he and the vice-chairman planned to testify at the hearings on amendments to the Water Resources Planning Act on July 10 and 11. Dr. Richard Dowd (Connecticut) advised that NERBC's new OCS Task Force had held its first meeting. After Dowd's report, Glen Kumekawa, Rhode Island's official representative, took Interior and EPA to task for public statements by their regional administrators to the effect that a New England regional approach to offshore development is impossible. New Hampshire's James E. Minnock wanted it clearly understood that his state supported involvement by NERBC in energy matters only if obstructionist and delaying tactics were avoided.

Turning to program and budgetary matters, the commission was faced with the problem of a changed federal fiscal year. Separate state and federal budget cycles would have to prevail until the states could supply a fifteen-month appropriation to achieve proper transition. A decision was made to terminate the Connecticut River Basin program after October 31, 1975 when the supplemental study was complete. Commission members were supportive of the proposed Lake Champlain Level B study, particularly under the new federal procedures calling for centralized funding through NERBC. A Plan to Study would be requested of WRC. Formal resolutions were then adopted establishing budget levels for FY 1976 and FY 1977.

The commission's regular quarterly meeting was held in September in Waterville Valley, New Hampshire. It was reported that vice-chairman Ristau had resigned, and Bernard Johnson of the Vermont State Planning Office would complete his term.

On NERBC program matters, chairman Gregg advised that LISS had been completed. The SENE study was nearing completion, and the Connecticut River Basin program was at the draft report stage. The Maine Guide Plan was now ready for commission approval. Warren Fairchild, director of WRC, reported that funds were available on a three-to-one matching basis to help the commission develop its CCJP, a statutory requirement of the agency.

Extensive consideration was given to the status of state coastal zone planning as reviewed by Michael Payne of the Department of Commerce's Office of Coastal Zone Management. Payne reported a positive feeling on the part of his office toward the use of regional agencies for coordination purposes. It was learned that the Federal Regional Council had created a sub-task force of its Task Force on Energy Development to provide a counterpart to the NERBC's purely state coastal zone task force. Chairman Gregg expressed regret at the formation of still another regional mechanism and suggested that a joint NERBC-FRC committee might head off criticism from state officials concerning excessive regional coordination.

NERBC was also advised of a new source of funding for staff studies — the Resource and Land Investigations (RALI) program administered by the U.S. Geological Survey. Support was reportedly available for the development of a methodology for siting onshore facilities. Following discussion, a study plan was approved for submission to the Geological Survey.

The final agenda item was a presentation by former Maine Gov. Kenneth M. Curtis of his recommendations on regional institutional arrangements for New England, a report described by Vermont's Arthur Ristau as the most significant such document in a decade. Curtis favored a merger of the NERCOM and NERBC natural resources planning activities, an expansion of NERBC's authority to include natural resource issues generally, and continued selection of the vice-chairman from the same state as the NERCOM cochairman. The Curtis report was scheduled to be presented to a future meeting of the New England governors, chairman Gregg advised. New Hampshire's Planning Director George E. McAvoy, a regular dissenter from NERBC actions, observed that his state was impatient with the large number of regional agencies. He favored a single umbrella agency covering all functions. Ristau spoke of the need for creative tension in any effective organization and felt that the NERBC should be challenged more. Discussion then ensued about the desirability of mixing state and federal representatives in a regional agency. HUD's Sheldon Gilbert observed that the two groups rarely display a separate identity and considered the joint representation actually a strength.

At the December meeting of the commission, chairman Gregg introduced Ben Yamagata, staff counsel to the Church subcommittee, who reported that the oversight hearings on the Water Resources Planning Act had been completed and that a summary of the proceedings was in preparation. The schedule was to submit a subcommittee bill to the Congress in the spring of 1976. Recommendations for strengthening the Title II commissions might well include direct control over planning funds, administration by them of the Title III state planning grants, and generally more "muscle" for the chairmen to achieve coordination. He expressed himself as much impressed with NERBC's strategies for natural resource decision making. Yamagata attributed Congress's lack of attention to river basin commissions to their small number nationally and their incomplete coverage of the country. He noted that the Southeast Basins Interagency Committee, a federal-state coordinating body for the past twenty years, was now seriously considering conversion to a river basins commission.

Turning to program matters, the commission learned from its flood plain task force that federal flood insurance, in some instances, was actually encouraging development on barrier beaches. One such case was in Rhode Island, Daniel Varin reported. Findings would be submitted to HUD in order to bring about reforms in the flood insurance program.

BOR regional director Maurice D. Arnold could find no interest among state liaison officers in the formation of a recreation committee, so the idea had been dropped, staff director Brown advised. Rudolph Hardy reported that an assessment of regional research priorities had been authorized by the New England Council of Water Center Directors which would relate well to NERBC's own regional assessment program. The prospects for a Lake Champlain Level B study appeared so promising that the staff had moved to identify prospective coordinating committee and study team members, the chairman advised.

Turning to the matter of funds available for CCJP activities, Gregg felt that time was too short to complete a full priorities report by the June 30, 1976 WRC deadline. He suggested a model report for a single basin and appointed a federal-state committee for this purpose. Priorities lists would be expected for five-year time intervals after the process got underway.

The commission then received written and oral reports on its RALI and coastal zone projects. The interim report covering potential on-shore impacts from OCS development was quite reassuring. Most of the initial impacts were likely to occur in established port areas. RALI project director Vincent Ciampa observed that much of the information assembled was not new but that it was rarely available to the public. As for the coastal zone task force, staff director Robert Brown reported that the state coastal zone directors had recommended a six-point, cooperative program. Much discussion ensued about one of the proposals, a documentary film describing what actually happens when OCS development takes place. NERBC members were concerned that the film be factual, informational, present both sides fairly, and not draw conclusions. Another element, a proposed Regional Technical Information Service Center, drew strong support from the state coastal zone officials. The proposal for a regional port development study attracted a positive response from Joseph Ignazio of the Corps of Engineers, but a warning about the costs of such an investigation.

1976

As the Commission gathered in Boston for its March quarterly meeting, the prospect of offshore oil development loomed even larger. Interior's Frank Basile, head of the Bureau of Land Management's leasing office in New York, reported that a Georges Bank lease sale could take place as early as late 1976 or early 1977. J. R. Jackson of Exxon advised the Commission that the American Petroleum Institute had decided against sponsoring a New England study of onshore impacts, feeling that its similar study for the mid-Atlantic region would suffice. Massachusetts's Evelyn Murphy had succeeded to the vice-chairmanship, and with state contributions from Maine and New Hampshire chronically in arrears, her first order of business would be to discuss the whole matter of state support with NEGC chairman Michael Dukakis.

In program matters, chairman Gregg reported that hearings had been held on the final report of the Connecticut River supplemental study. This project was essentially complete. He advised that the NERBC task forces on flood plain management, coastal zone, and OCS had been very active. Of particular interest were the pending amendments to the coastal zone management act that would provide special Section 308 grants to the states to compensate for OCS-related impacts, and would require federal consistency "to the maximum extent possible"[80] with state coastal zone plans. Another provision would offer 90 percent federal funding to encourage interstate coordination. Vice-chairman Murphy reported that the New England governors had also agreed to form a regional group, which she would chair, to prepare for an expanded New England fishing industry upon passage of the pending 200-mile fisheries jurisdiction bill. She predicted almost certain conflict between oil and fishing interests on Georges Bank.

Despite the wind-down of its comprehensive studies, NERBC continued to remain active in water resources related planning. Maine and Massachusetts had agreed to become pilot states to explore the consolidation of Section 208, Title 701, and coastal zone planning programs. A Proposal to Study had been authorized for the Lake Memphremagog region in northern New Hampshire. Commission staff and state personnel from New York and Vermont were monitoring closely the International Joint Commission's activities at Lake Champlain where requests for proposals had been issued for its projected studies. Additionally, NERBC had approved a study of the hydro power potential of small dams in New England, a sequel to the investigation of large sites conducted recently by the Corps of Engineers, for inclusion in its FY 1977 and FY 1978 budgets. Interest was also expressed in the nonstructural flood control alternatives provided by Section 73 of the Public Works Act of 1974. Commission members gave high priority to case studies of this approach within the Connecticut River Valley and had identified six possible sites.

Chairman Gregg then directed the commission's attention to the matter of the CCJP. The members approved a suggested approach to a first-cut priorities report and the CCJP outline required by WRC to qualify for increased federal support. The commission also supported the chairman's desire to expand the membership of the CCJP committee to include a representative of each state. If the priorities and other documents achieved their goal of redirecting water project investments, then the CCJP committee would become a central element of the entire NERBC program.

The May meeting of the commission was held in East Hartford, Connecticut and was preceded by a field inspection of the flood control dikes protecting this portion of the Connecticut Valley. Following the usual round of caucuses and committee sessions, members were advised that the Curtis report had been received favorably by the governors and that a special committee had been appointed to improve cooperative relations between NERCOM and NERBC. A draft resolution was reviewed and approved, which would make the governors' special representative for NERCOM automatically the vice-chairman of NERBC, would provide for an annual review of NERBC's program and budget by the governors, and would explore further steps toward consolidation through the route of changes in legislation or presidential executive orders. Commission members felt that the respective missions of the two regional agencies should be clarified as well.

After approving a number of appointments, the commission turned its attention to the priorities report for FY 1976 recommended by the CCJP committee. A number of adjustments were made, projects relating to hydro power and groundwater investigations receiving an elevated position in the priority rankings. The proposed diversions at Northfield Mountain and the Tully-Millers Rivers were deleted from the list of new construction starts, but Joseph Ignazio of the Corps of Engineers insisted that a narrative section be included setting forth the history of the two proposals and the positions of the affected states and federal agencies.

In budgetary action, a program and budget for FY 1977 and 1978 were approved, including the first steps toward a regional port study.

For its September meeting, the commission traveled to Sebasco, Maine. In accordance with the New England governors' resolution,

Massachusetts's special representative for NERCOM, Thomas J. Doyle, had become the new vice-chairman.

In program business, the commission agreed to accept supplemental funds and participate in a series of national workshops designed to transfer the onshore facilities siting technology developed during the RALI project. This was followed by a presentation and roundtable discussion of the three principal reports produced by this project: a factbook on OCS leasing procedures, preliminary estimates of onshore facility impacts, and a planning methodology for locating such facilities. A special guest was Derek Lyddon, Chief Planner for the Scottish Development Department, and a veteran of such activities in the North Sea region. The NERBC/RALI group had concluded that finding appropriate landfalls for natural gas pipelines was likely to be the most immediate decision for the coastal states. Vincent Ciampa, RALI project director, reported that the next phase would involve the application of the three reports to the New England states on a case-by-case basis.

At the December meeting in Boston, Gov. Michael Dukakis told the commission that there had been a warm response by the governors to the idea of strengthening relationships between NERCOM and NERBC. As tangible evidence of the new spirit of cooperation, NERBC invited NERCOM's program directors to outline their current activities in economic development, labor, transportation and energy. An obvious point of commonality was the onshore facility siting activity of the commission's OCS, coastal zone, and RALI working groups. Interior's Roger Babb noted how impressed he had been with the professionalism displayed at the recent environmental impact hearings on the proposed Georges Bank lease sale.

Turning to commission matters, chairman Gregg advised that OMB had recommended the elimination of all funding for WRC (Title I) and the state planning grant (Title III) programs, but the incoming Carter administration might well reverse these decisions. Gregg also reported that he had asked the CCJP committee to exercise an additional responsibility in reviewing the commission's annual program and budget for FY 1979. NERBC members were also advised that a preliminary study of port activities and needs conducted by former LISS project manager David Burack had yielded promising results.

The last item of business was the review and approval of the final report of the Connecticut River Basin supplemental study. The report itself was acceptable, but the summary was not. The Corps of Engineers, in particular, objected to the proposed substitution of the report for the commission's earlier NERBC 1980 Basin Plan, which still called for structural measures to relieve flood problems in portions of the Connecticut Valley. After much discussion, the main report was approved for publication and transmittal as required by law, but the basin plan was not amended.

1977

Charles M. McCollum, NERBC's new vice-chairman, opened the March meeting in Boston by reporting that the state members had agreed to seek the full $5 million funding authorized for the Title III state grants program. The incoming Carter administration was expected to be more sympathetic than the Ford administration. He noted that three of the present state members were also serving as alternates to NERCOM, a fine beginning to the process of coordination favored by the governors.

Progress reports were received on the NERBC's Connecticut River Basin and flood plain management studies. The Section 73 investigations on the Connecticut River were bringing these two programs closer together. The next step for the regional ports study would be a meeting of port authorities and state officials to discuss needs. Full sets of SENE maps were now available. The proposed hydro power study was reported safely in the President's FY 1978 budget. NERBC was also encouraged to consider other water-related projects. Involvement in the national wild and scenic rivers study was urged by Maurice D. Arnold, regional director of the Bureau of Outdoor Recreation. The U.S. Forest Service's Wendell Doty and Yale forestry dean Charles H. W. Foster recommended commission participation in the regional resurvey and planning scheduled for New England's extensive forest resource. NERBC had also been asked to assume responsibility for the EPA-funded public participation program relating to water quality. Commission members, however, declined to take part in ventures they regarded as peripheral to their central mission.

The commission then set to work on the FY 1978 and 1979 programs and budget. Members felt that the basin program should have precedence over special studies, which should be supported entirely by outside funds. Of first priority in special studies would be matters referred to the commission by the governors. The next priority should go to issues related to the central mission of the commission. The underlying concern was the need to develop an effective constituency. This could be done only with a narrower focus to the program. Any more detailed budget review was deferred to an April 20 meeting of the CCJP committee.

The June meeting of the commission was held at Hyannis, Massachusetts. It was reported that the New England port operators had received favorably the idea of a regional ports study and were prepared to participate actively in it. Vice-chairman McCollum insisted that NERCOM also be invited to play a role in the study. A suggestion for citizen participation in the Connecticut River Basin coordinating group met with mixed reactions, but the commission agreed in principle if the coordinating group could suggest an equitable method.

Colonel John Chandler, the Corps of Engineers' NED engineer, then spoke of his agency's difficulties in locating areas for the disposal of dredged spoil. He urged that NERBC view this as a significant regional problem and consider using the Long Island Sound area as a pilot project to work out approaches. The matter was received favorably and referred to the NERBC staff for further investigation.

Budget and program recommendations for FY 1978 and 1979 were received from the CCJP Committee and approved.

The remainder of the session was devoted to a special mini-conference on federal water policies and programs. This was sparked by President Carter's recent environmental message to Congress announcing a national water policy review. Several members had also attended the National Conference on Water held at St. Louis in May where Secretary of the Interior Cecil Andrus had confirmed the President's intentions. Panelists included Carol Jolly of the National League of Women Voters, William Miller of the Water Resources Congress, and Gary Cobb, acting director of WRC. The New England participants were not reticent in expressing their views. NERBC's staff director, Robert Brown, attempted to summarize the proceedings. There should be sufficient flexibility in national water policy to recognize regional differences and achieve integration of programs at all levels—possibly through a comprehensive water grant program. Federal water policy should be implemented at state, substate, and local levels using existing institutions and involving those affected. New Englanders were particularly partial to the use of nonstructural alternatives. They gave highest priority to problems of dredged spoil disposal, eutrophication of lakes, and water quality degradation. Chairman Gregg closed the session by stating that a summary of the proceedings would be made available to the President's national water policy review committee.

At the September meeting at Vergennes, Vermont, NERBC was briefed on the progress of the Lake Champlain Level B study and approved the regional assessment report for transmittal to WRC. It received reports of the Connecticut River Section 73 study and the coastal zone liaison program, and was brought up to date on the status of the hydro power investigation. It also received a briefing on current energy projects from Richard RisCassi of NERCOM's energy office.

Members then moved to accept the recommendations of their Connecticut River Basin group. One citizen would be selected by each basin state member and added to the coordinating body. They also approved the summary of the final report and authorized amendment of the 1980 Connecticut Basin Plan to conform with its recommendations.

Three new RALI projects appeared likely of funding: natural resources data management, regional power systems planning, and urban area water management. The commission approved the development of Plans to Study in each instance.

NERBC members then returned to the subject of dredged spoil disposal. A Dredge Management Committee was appointed by the chairman consisting of representatives from New York, Connecticut, and the Corps of Engineers with Massachusetts and Rhode Island participating as observers. The commission was not responsive to Chandler's suggestion that NERBC examine the regional implications of the Dickey-Lincoln School project in northern Maine, a controversial flood control and hydroelectric project authorized for Corps study. Maine was less than enthusiastic about NERBC involvement, and the commission was reminded that the governors had expressly decided not to refer the project to NERBC for evaluation. Recognizing the sensitivity of the situation, chairman Gregg stated that he would consult with vice-chairman McCollum before responding. He urged the state members to consult with their governors.

Two special discussion sessions were then held. The first centered on the report, *A Look at Lakes,* prepared by the New England Council of Water Center Directors, the final report of its study of the effects of urbanization on such water bodies.

The second, *Refining Federal Water Policies To Meet New England Needs,* was an outgrowth of the discussions on Cape Cod and the July 28–29 Boston hearing of the President's water policy review team. A draft report had been prepared and circulated by the NERBC staff outlining a possible New England position. Chairman Gregg observed that NERBC could not adopt a position on national water policy because of the presence of federal members, but he saw no reason why the states could not develop a position of their own if they so desired. All but New Hampshire concurred, and Massachusett's Evelyn Murphy agreed to serve as chairman of an ad hoc state members' committee for this purpose.

The December meeting of the commission was held in Boston. The Corps of Engineers reported that, under the President's dam safety initiative, nearly 1,000 high risk dams had been identified in the New England region alone. Its Section 73 studies in the Connecticut valley were reported as proceeding in fully coordinated fashion with those of the NERBC.

In other matters, the commission approved a study of the environmental impacts of offshore oil and gas pipeline construction and operation requested by EPA. It was further advised that NERBC's assistance to EPA in public participation would continue for another year.

At the close of the session, chairman Gregg introduced representatives of four other river basin commissions, who had agreed to advise the New England–New York region in the development of a consolidated position on national water policy. Gregg reported that seven national task forces were currently at work. Secretary Andrus was scheduled to appear before the National Governors Association conference in January. In the meantime, its Subcommittee on Water Management had expressed interest in using the New England–New York statement as a model for its own position paper. From the reports of the visitors it was apparent that there was general agreement with the Northeast's thrust toward a strong role for the states, and particular interest in the need for respect for and flexibility toward regional differences. However, the visitors felt that New England's emphasis on conservation as the cornerstone of federal water policy might not fit many parts of the West. They favored more emphasis on water planning to be carried out by the states with federal inputs. It was also pointed out that river basin commissions could not play a central role in federal water policy since they were not uniformly established throughout the country. Regional differences were apparent in the concern of some commissions over water rights, institutional arrangements, and cost sharing for water development projects. The Western commissions, in particular, were adamant about maintaining the rights of the states to manage water resources without federal interference.

1978

The March meeting of the commission was the first since its inception without Frank Gregg. It was presided over by Col. John Chandler of the Corps of Engineers, named by Gregg as alternate chairman shortly before his departure for Washington to assume the position of director of the Bureau of Land Management. A new vice-chairman was also in office—Sydney Frink of New Hampshire, Gov. Meldrim Thomson's representative for both NERCOM and NERBC. In the audience but still without portfolio was Dr. John R. Ehrenfeld of Energy Resources Co., a Boston area environmental consulting firm, who had been named by President Jimmy Carter as chairman designate of NERBC. With Washington still in transition, the commission turned to visitor John Roose of WRC for an explanation of current developments. He reported much criticism of WRC and the river basin commissions but felt that any resolution of their status must await the disposition of the President's recommendations for changes in national water policy.

Vice-chairman Frink reported that the states had agreed in caucus to work for restoration of budget cuts. He would testify at the reauthorization hearings in Washington in April. Rhode Island's Daniel Varin had agreed to monitor the federal natural resources reorganization proposals. Frink also reported the states as willing to consider a 5 percent increase in state appropriations to help offset inflationary costs in the NERBC's operating budget.

Two other significant developments were reported at the national level. Maurice D. Arnold reported that his agency, the Bureau of Outdoor Recreation, had been reorganized into the new Heritage, Conservation and Recreation Service, whose responsibilities would include the protection of cultural and environmental resources generally. And Charles Dingle of the Department of Agriculture reported passage of the Resources Conservation Act of 1977, which would require the Soil Conservation Service to engage in program assessment and planning on a five-year timetable.

The commission then turned to the 1980–84 Priorities Report, developed by the staff and reviewed by the CCJP committee. Candidate projects were grouped under five categories: data collection, research, planning, feasibility studies, and implementation. Numerical rankings were assigned each project based upon problem severity, population affected, national significance, regionwide significance, and degree of support. Although a measure of concern was expressed over the lack of detail in the listings, the report was approved for transmittal to WRC. The NERBC meeting then moved to a special workshop on coastal flooding, inspired in part by New England's severe coastal storm of February 6–7, 1978. Approximately one-hundred participants heard comments from nineteen specialists representing federal, state, and local levels. Among conclusions reached was the need for prompt land acquisition to secure flood-prone areas against rebuilding. Full documentation by aerial photography was also urged, as well as a complete assessment of the social, economic, and environmental damage. NERBC was encouraged to develop a special planning program for the coastline, which would include improved maps of flood-prone areas. NERBC should also serve as a facilitator for an inventory of the physical effects of the storm on coastal land forms.

The June quarterly meeting was held in Newport, Rhode Island, with Gov. J. Joseph Garrahy as the featured speaker. In a brief ceremony, Interior Assistant Secretary Guy R. Martin installed John Ehrenfeld as the new chairman. Martin had been pressed into service earlier to explain the President's new water policy at separate meetings of federal and state members. At both caucuses, concern was voiced over the proliferating committee assignments and the lack of funds for effective participation. Some form of committee consolidation was clearly in order.

In other matters, the commission approved a program and budget for FY 1979 and 1980 as recommended by its CCJP committee. It also approved a regional policy statement of flood plain management prepared by its special flood plain task force and adopted it as an element of the CCJP. The revised Merrimack Basin Overview was approved, as was a proposal by Massachusetts that the residue of the contributions for the Connecticut River Basin Program be spent for public education on the issues of flood plain management. The commission learned of an Interior urban recreation study designed to assess the adequacy of open space and the delivery of recreation services in urban areas.

A special panel then assembled to describe Rhode Island's newly approved coastal zone management program, one of the first in the nation. Participants included John Lyons, chairman of the Rhode Island Coastal Resources Management Council, the state's new management and regulatory body. The council had come into being following a massive planning and education program by a citizen group appointed by former Gov. Frank Licht eight years earlier. Steven Olsen, coordinator of the University of Rhode Island's Water Resources Research Center, then described the structure and responsibilities of the state's coastal zone plan, which covered all activities in Rhode Island tidal waters. He was assisted by Malcolm E. Graf, former staff director of the NERBC and engineering consultant to the Council, who explained the significance of the consistency provision contained in the federal coastal zone management act. For a small state with extensive federal installations, these provisions are crucial, Graf observed.

The September meeting directed the commission northward to Portsmouth, New Hampshire. Here, vice-chairman Sydney Frink expressed state members' concerns over the lack of private sector involvement in the affairs of the NERBC and suggested that the CCJP committee develop remedial recommendations for the December meeting. There was also growing uneasiness over the implementation of the President's new water policies. State representatives hoped that any new Water Resources Council regulations would preserve the measure of flexibility enjoyed by regions such as New England. This set the stage for a full report on Washington activities by Dr. Leo Eisel, director of WRC, who described the current disagreements between the President and the Congress on the funding of water projects. For that reason, the public works appropriations bill was likely to be vetoed. In the meantime, President Carter had instructed WRC to proceed with a revision of its principles and standards document to emphasize nonstructural alternatives and conservation approaches. The President wanted the council to undertake an independent review of Corps, Agriculture, TVA, and Reclamation projects to be certain that they meet the new criteria. Eisel reported that the state grants program (Title III) was likely to go from $3 million to $25 million in the next Carter budget.

In program matters, two planning reports were presented to NERBC: the Kennebec Basin Overview (the second in the series) and a new slide presentation entitled *Lake Champlain: Time of Choice*. A progress report on the commission's hydro power study revealed that existing dam sites could supply 15 percent of the additional generating capacity needed by the region over the next ten years. Agreements with the Geological Survey were approved activating supplemental RALI studies on OCS information maintenance, natural resources data management, urban water conservation, power plant siting information, and coastal flooding. The chairman was authorized to publish for ninety day review an amendment to the 1980 Connecticut Basin Plan terming the proposed Northfield diversion an alternative of "last resort."[81] This would make the document consistent with Massachusetts's present policy position.

A special session entitled *Focus on the Piscataqua* was arranged by vice-chairman Frink to illustrate the opportunities and constraints on economic development faced by an interstate river.

At the December meeting of the commission in Boston chairman Ehrenfeld reported a number of developments at the Washington level. OMB had requested WRC to undertake an evaluation of the effectiveness of Level B planning with special attention to implementation. NERBC was expected to play a central role in that study. Further, Ehrenfeld had provided leadership for all six river basin commissions in developing uniform priorities reports, procedures, and formats. He requested and received formal commission approval for the use of these guidelines.

In program matters, the next two basin overviews (the Thames and Penobscot rivers) were reported to be in draft form. The hydro power inventory now contained a listing of 10,000 New England dams. Lawrence Bergen of the Corps reported that the February, 1978 storm had caused $300 million worth of damage in forty-six coastal communities. A preliminary report of the New England Energy Congress was available, setting forth an energy agenda for the New England congressional delegation. Bartlett Hague of EPA announced a recent EPA/Interior conference on water cleanup and the land, which had resulted in a national, interdepartmental agreement governing joint activities in regions such as New England.

The commission was advised of a new initiative by the U.S. Geological Survey to assess regional aquifer systems. The chairman was instructed to recommend a study of New England's saturated stratified drift aquifers. An equal amount of interest was displayed in cosponsoring with the Geological Survey a special workshop on coastal mapping techniques. Commission members also felt that NERBC's Housatonic and Lake Champlain basins might qualify for inclusion in the Survey's program of river quality assessment.

Left to the end was a discussion of the commission's program objectives, specifically requested by chairman Ehrenfeld to provide insights for him in administration and management. Comments were received on all facets of the NERBC program. Of particular concern to members were matters relating to the CCJP and the consistency of federal programs with regional water plans. After considerable discussion, NERBC agreed in principle with the revised committee structure developed by the chairman and staff and authorized him to draw up the necessary by-law amendments. The commission's current twelve committees would be reduced to four: CCJP, basin planning and coordination, regionwide studies, and coastal programs. The chairman would chair them all. Left alone would be the present working-level structure of task forces, study teams, and work groups.

1979

NERBC's March meeting took place in Boston. Arthur Markos of Rhode Island, who had succeeded to the vice-chairmanship, reported a willingness of state members to testify at water resources and appropriations hearings in Washington in early April. Because of uncertainties in funding, New York would continue to participate in NERBC activities but would abstain from voting on all matters.

The commission was advised of a proposed merger of the New England Division of the Corps of Engineers with the North Atlantic Division, a move that most members felt would be to the region's disadvantage. Chairman Ehrenfeld also noted a forthcoming water policy workshop, one of four nationally sponsored by the NERBC and the Department of the Interior, to gauge progress in meeting President Carter's water policy initiatives.

In program matters, the Thames Basin Overview was approved for publication and the Penobscot and Kennebec overviews were reported as nearing completion. A draft interim plan had been prepared for the disposal of dredged material in Long Island Sound, but NERBC members wanted more time for review before authorizing its distribution for public comment. The services of NERBC as mediator seemed to have worked well. It was further reported that the states had agreed to contribute $4,000 each to continue the coastal zone liaison service.

The commission next turned its attention to FY 1981. It approved an updated priorities report for the five-year period 1981–85, then approved a revised FY 1980 and FY 1981 program and operating budget. Chairman Ehrenfeld adjourned the meeting to a special workshop on strategies for state participation in OCS exploration decisions, an outgrowth of NERBC's previous work under the RALI program.

The June session of the commission was held in Hartford, Connecticut. At the earlier caucuses of state and federal members, the NERBC had learned about new water legislation filed by Sen. Daniel Moynihan (New York), which would provide a pivotal role for the states in federal water project authorization. The members had also been briefed on the northeast water study being undertaken by the Nova Institute and the Northeast-Midwest Institute, under sponsorship of a coalition of Congressmen from the "frostbelt" section of the country.

In program matters, members received an extensive briefing on the Housatonic Basin from staff members engaged in the basin overview study and from the Corps's urban waters staff. The parallel studies appeared to be well coordinated. Less satisfactory were the dual Section 73 studies on the Connecticut River, where the NERBC element had been deferred pending the resolution of work plan problems. The interim dredge management plan for Long Island Sound was cleared for public release. The commission also authorized the submission of proposals to explore the potential and strategies for achieving water conservation in the region, a subject of concern to many urbanizing areas. If accepted by WRC, the study would require a 25 percent contribution in cash or kind from the member states.

Hugo Thomas of Connecticut used the occasion to press for still further revisions in the 1980 Connecticut Basin Plan. Upon his motion, the commission resolved to modify the plan to recognize the potential problem of radioactive releases from the Vernon nuclear plant in Vermont and to ensure that riparians upstream from a point of diversion would not be required to pay for advanced waste treatment in order to meet water supply standards. Commission members found the modeling study of Connecticut River water demands conducted by Resource Analysis Inc. generally inconclusive and accepted the cumulative impacts report without adopting its recommendations.

The formal NERBC meeting then adjourned to a day-long special conference on groundwater opportunities and needs cosponsored by the U.S. Geological Survey, the state of Connecticut, and the commission. It was clear from the discussions that this was a topic of growing interest and concern in many portions of New England.

The September meeting of the commission was held in Portland, Maine, and a special program focusing on Portland Harbor and Casco Bay was arranged for members and guests.

In the business meeting, Chairman Ehrenfeld reported that the river basin commissions nationally had met with OMB on budgets and planning procedures and he felt that some progress was made. He advised that no resolution was in sight of the conflict between President Carter and the Congress on the matters of independent project review and an independent chairmanship for WRC. He further reported that the General Accounting Office had initiated an analysis of river basin commission programs with NERBC's as one of three selected for special case studies. Staff director Robert Brown added that WRC had instituted a special task force to examine Level B planning procedures and plan utilization which would be chaired by John Ehrenfeld. The Federal Emergency Management Agency had become a new federal member of NERBC.

Staff program reports revealed that additional overviews were underway for the Piscataqua, St. Francis, and Saco River basins. The report of the Kennebec Basin Overview was reviewed and approved, the third such study completed in the series. Study manager Paul Vachon then presented a final report, *Shaping the Future of Lake Champlain*, and described its eleven major water resources management program recommendations. The report was accepted with enthusiasm by the commission. The New York and Vermont members urged that a follow-on project be considered for Lake Champlain Basin and possibly also the Kennebec River Basin.

NERBC also approved a detailed report from the CCJP Committee, which recommended revised program/budget review and priorities processes procedures, expanding the scope of the CCJP to the entire region, and enabling broader public involvement.

At its December meeting in Boston, commission attention was directed to several matters relating to coastal activities. The final report suggesting means of participation by the states in OCS exploration decisions was reviewed and approved. Included in the recommendations were individual state policy committees established under state law and improved coordination and involvement by agencies such as NEIWPCC, NERCOM and NERBC. The commission's Coastal Activities Committee reported that it had approved a special staff report, stemming from the severe winter storm of 1978, dealing with coastal hazards. It was felt that the existing state coastal zone management programs were in the best position to follow through on the recommendations. In other coastal matters, the Maritime Administration had expressed interest in helping fund the proposed regional ports study.

The CCJP Committee then asked and received concurrence with its proposed priorities ranking for the period 1982–86, an updating of the earlier document. Chairman Ehrenfeld recommended acceptance of funds from the University of Arizona to enable NERBC to participate in a nationwide effort to set up computer-based indices to water supply data, a program that would assist the commission in its own studies of water conservation. He also reported that the staff was prepared to cooperate with EPA for the third straight year in the area of public participation.

Commission members approved the Piscataqua Basin Overview and received a progress report on the urban water conservation study, including its initial two case study communities, which was proceeding under the guidance of NERBC's Region-wide Studies Committee.

1980

At the March meeting of NERBC in Boston, chairman Ehrenfeld reported a number of developments in Washington. The task force on Level B studies had completed its assignment for WRC, but the two others dealing with planning procedures and plan utilization and on plan review were still at work. He advised that there had been little progress toward a reauthorization of the Water Resources Planning

Act. Vermont's Timothy Hayward, the new vice-chairman, reported that the states had agreed to increase their contributions above base by 8 percent to offset inflation. State matching funds were expected to exceed $200,000 in FY 1981.

The basin overview program was described as progressing well. The commission approved the Piscataqua and New Hampshire Coastal River Basins Overview, which recommended a case study of Portsmouth Harbor as part of the regional ports study. It was reported that the overviews of the Saco and Southern Maine and of the Lake Memphremagog/St. Francis River Basins and the follow-on activity for the Connecticut River and Lake Champlain projects, were proceeding.

Robert McIntosh, regional director of Interior's Heritage, Conservation and Recreation Service, then invited cosponsorship by NERBC of a recreation rivers study for New England. The commission also learned that the draft interim plan for dredge management in Long Island Sound appeared to have survived all of its public review hurdles. The next step would be the development of a long-range dredge management plan for New England as a whole. It also learned that the urban water conservation project was ripening into an effort to inject water conservation measures into state water resources planning generally.

As usual, the CCJP Committee report covered most of the fiscal actions required by NERBC. Included were the necessary revisions in the FY 1981 budget and approvals for the program and budget for FY 1982. NERBC decided to renew its agreements with EPA and the state coastal zone officers for another year. Funds from the latter would be supplemented by special project funds from coastal states for research into required offshore discharge permits. It was likely that NEIWPCC would cooperate in these investigations.

The commission then directed its attention to the matter of constituency building discussed at earlier meetings. Adopted by consensus were a number of basic policies recommended by the CCJP Committee. They included expanded opportunities for nonmember participation by independent organizations and municipal officials.

The June meeting of the commission was held at Amherst, Massachusetts. Chairman Ehrenfeld said that he was prepared to begin improving nonmember participation by appointing a steering committee on public involvement and by streamlining the existing advisory apparatus in the Connecticut Basin, steps recommended by the CCJP Committee at the December meeting. He reported that the President's water policy initiatives were still not implemented, and, in the absence of a reauthorization bill, the fate of appropriations requests under titles I, II, and III was in doubt. The state members, in fact, had held an all-day meeting to discuss the current situation. Various individuals had agreed to help advance the New England cause before the National Governors Association and the appropriations committees of Congress.

In program action, NERBC accepted the Saco and Southern Maine Coastal Basin Overview for publication and distribution and formally adopted the interim plan for the disposal of dredged materials from Long Island Sound, amending the region's CCJP to include that provision. With Corps of Engineers' cooperation, the next phase of the dredge management project would take place in Narragansett Bay, it was agreed.

At a special briefing on acid rain, chairman Ehrenfeld noted the involvement of two regional groups, NEIWPCC and the Northeast States for Coordinated Air Use Management (NESCAUM). Of the states, New York was described as having the most active research program under way.

In committee reports, members learned of a successful flood plain management conference in Westfield, Massachusetts and an imaginative urban water conservation workshop in Amherst, both of which NERBC had sponsored.

The final item of business was a change in the commission's bylaws to conform with the new federal fiscal year, requiring approval of a special budget for the transition quarter of July 1–September 30, 1980.

The September meeting of the commission was held in West Lebanon, New Hampshire. There was good news from Washington—a federal budget had finally been enacted by the Congress. On the states' side, acting vice-chairman Bernard Johnson (Vermont) reported a desire by the members to update the New England–New York consolidated water policy statement and to bring the revisions before the National Governors Association's winter meeting.

The commission received a progress report on the Androscoggin River Basin Overview and accepted the Housatonic Basin Overview for publication and distribution. Implementation of the Lake Champlain Level B study had been accelerated by a working agreement with the Department of Agriculture's Soil Conservation Service with special attention to shoreline management considerations. NERBC's ports and harbors study was reported to be at the stage of regional framework analysis, and RALI-supported projects were well under way. The dredge management investigation of Upper Narragansett Bay had completed its initial site-screening phase.

Among other items, Maine's Craig TenBroech reported the recent discovery of a large sulphur-zinc-copper deposit in the Bald Mountain section of northern Maine. Mineral extractions could become an important topic for NERBC's regionwide studies program, he observed. The commission's hydro power project had reached the stage of workshop analysis to help formulate development strategies. The required public hearing was being held in conjunction with the commission's September meeting. A preliminary plan of study had been prepared for an expansion of the urban water conservation program into a series of statewide programs. Chairman Ehrenfeld also advised the commission that a first meeting of the new steering committee on public involvement had been held recently in Boston with fifteen of the twenty members present. The discussions were candid and to the point, he observed.

The September meeting marked the beginning of a new effort by the commission to update members regularly on state and federal activities. The Corps of Engineers and the U.S. Geological Survey were the first two agencies to present overviews of their programs.

The final meeting of 1980 occurred in December in Boston. As in so many other years, the Congress was reported to have adjourned without reauthorizing the Water Resources Planning Act. However, it had taken action to continue Titles I, II, and III for another year. With a new administration assured by the November elections, the state of uncertainty was likely to continue for some time.

In program matters, a priorities document for 1983–87 was reviewed and approved by the commission upon recommendation of its CCJP Committee, as was a concepts paper designed to provide an organizational framework for the future work of the CCJP. The CCJP Committee agenda items were gradually becoming the central focus of the commission's activities.

The basin overview for the Lake Memphremagog/St. Francis River region was next reviewed and accepted. Also presented was a preliminary overview for the Connecticut Coastal Basins. NERBC learned of plan and program implementation at Lake Champlain and within the Connecticut River Basin, where citizen and education activity continued to run high. The ports and harbors study was reported as essentially 75 percent complete, and the long-range aspects of the dredge management study were about to get under way. NERBC's coastal activities staff observed that on this first anniversary of the initial Georges Bank lease sale, the Geological Survey had finally deemed complete the exploration plan submitted by Exxon.

In other program matters, the proposed state water supply conservation project had reached the stage of a detailed plan of study. The commission's hydro power study would be completed in December

1981. The regional workshop held in Lebanon, New Hampshire in September had been oversubscribed, such was the public interest in the topic.

The elections of November 1980 confirmed the suspected trend to conservatism throughout the country. For environmentalists accustomed to the cordiality of the Carter administration, Reaganism was a matter of real concern. A Western President could be expected to return Interior to its historic preoccupation with the development of resources on the public lands to the detriment of the urban Northeast. If rumors were correct James F. Watt, president of the Mountain States Legal Foundation and a fervent supporter of the Sagebrush Rebellion, would be the next Secretary of the Interior. As a former director of Interior's Bureau of Outdoor Recreation, Watt knew where the bodies were buried. He was reportedly not an admirer of the river basin commissions, although it was said that he regarded NERBC as the best of the lot. At the least, it was likely that wholesale changes would accompany the advent of the Reagan administration, and extensive surgery would have to be performed on the federal budget if the President-elect's campaign promises were to be kept. Although some were apprehensive about the future, most bureaucrats were merely curious. The inertia of government was simply irresistible, and somebody else's agency or program would be the one affected.

Nonetheless, the ax fell on NERBC chairman John Ehrenfeld shortly after the first of the year. He informed his colleagues in a memorandum that he had received a letter from President Reagan terminating his chairmanship effective February 17, 1981. Acting under the vague provisions of Section 203(c) of the Water Resources Planning Act, Ehrenfeld designated First District U.S. Coast Guard Comdr. Stephen L. Richmond, a man with considerable experience in weathering stormy seas, as alternate chairman. He also established a Staff Transition Committee under the direction of Staff Director Robert D. Brown to carry on the day-to-day business of the commission until a new chairman was appointed. In a farewell letter to his associates, Ehrenfeld spoke with warmth of his NERBC experience. But he took the occasion to offer a warning about the future. Planning, he said, is never the most popular of government functions and is generally one of the first targets in times of fiscal austerity. To abandon water resource planning now would be a costly mistake.

We need better public and private decisions guided by and, in some critical areas, constrained by a dynamic planning process. The plans and policies that will result from the process must evolve from a careful examination of the consequences, both positive and negative, of alternate paths, must view the world as a finite, interrelated, complex system, and must reflect public values set far from the parochial arenas of individual, localized disputes.[82]

1981

On March 10, 1981 the national storm clouds broke over the water resources community. The new administration's revisions of the FY 1982 federal budget had eliminated funds entirely for all programs administered by WRC. The Administration had concluded that "the commissions do not perform any function or provide a service the States are not able to accomplish themselves."[83] With Secretary Watt likely to chair WRC in addition to his leadership of the Reagan natural resources cabinet, Interior's pronouncements carried particular significance.

The commission's first quarterly meeting of 1981 was held in Boston in March. Only two official NERBC members were present: the representatives from Vermont and the Department of Agriculture. The remainder were alternates. Maine and the Department of Housing and Human Service were not represented at all. Of the six interstate agencies, only the Interstate Sanitation Commission and the Merrimack River Valley Flood Control Commission were represented—both by alternates. The afternoon of the first day was devoted to a special panel session on drought in the Northeast.

Commander Richmond opened the official business meeting on the second day with a confirmation that NERBC had been zero-budgeted effective September 30, 1981. He reported a special CCJP Committee meeting the evening before to devise an orderly phase-out plan, the main thrust of which would be to produce worthwhile, usable, end products within the time specified, yet retain some core funds and staff to carry into FY 1982 if the situation should change. Richmond observed that water resource management problems would continue to exist in New England with or without the commission. The best the agency could do would be to pass on to the next generation of managers its wealth of information and experience, he said. The alternate chairman planned to write the governors and the Department of the Interior about the commission's accomplishments and would suggest ways by which a program might be continued. He wished it to be a matter of record that the staff had given extra hours under less than pleasant conditions, and he congratulated them for being willing to finish off programs and produce a meaningful final document under such circumstances. Commander Richmond added that he had been in touch with Interior Secretary Watt's office and could learn only that the water resources situation still remained in a high state of flux.

Despite these ominous signals from Washington, the commission continued with the regular business at hand. It voted to accept the Androscoggin River Basin Overview as presented; approved a revised FY 1982 and a FY 1983 program and budget on a contingency basis; and extended routine agreements with the state of Maine and coastal zone management officials for staff and liaison services. A brief update was presented on the National Weather Service's flash flood warning program. The commission also took action to authorize its CCJP Committee to review and release a regional policy statement on acid deposition.

Alternate vice-chairman David Neville gave the states' report on behalf of New Hampshire's Ronald Poltak, the new vice-chairman. He advised that Robert Hook and William Horne of the New York Department of Environmental Conservation had proposed the formation of a successor Northeastern states water council to bring the common interests and concerns of state water managers to the attention of the federal government. This informal coalition would be the counterpart of a similar group representing the western states. Massachusetts's Elizabeth Kline agreed to keep state members advised of any such development. At a brief caucus following the CCJP Committee meeting, a majority of the state members had instructed vice-chairman Poltak to write Secretary Watt and others urging retention of Titles I, II and III programs.

The bulk of the remaining time was devoted to the CCJP Committee's proposal for terminating the commission's program. The members and alternates reviewed the status of all remaining studies one by one. By good fortune, only the long-range dredge management and the water supply/conservation programs would require substantial foreshortening. In formal action, NERBC approved the CCJP Committee's phase-out plan, delegated authority to the alternate chairman to carry it out, and further delegated authority to the Committee to amend the plan if conditions changed. The session then adjourned; the prospect of the next quarterly meeting remained conjectural.

On March 17, Interior's Russell informed Commander Richmond that Secretary Watt had indeed been appointed chairman of WRC by the President, that the status of WRC itself was indeterminate, that Watt was interested in setting up a new Office of National Water Policy in his own department with a staff of twenty-five and a budget of $5 million, and that the Secretary was anxious to proceed quickly. Russell reported some congressional opposition, however. As an example, Senators Jennings Randolph and Robert Stafford, the two ranking members of the Senate Committee on Environment and Public Works, had written the Senate Appropriations Committee urging continued, if reduced, funding for the Water Resources Council, the river basin commissions, and state water resources planning.

On March 27, WRC director Gerald Seinwill met with his new chairman. Watt was quite specific in setting forth his views. There would be no formal council meetings during the year. The Title III grant program to the states must go. No new river basin commission chairmen would be appointed, with the possible exception of the Upper Mississippi Commission, which was nearing the end of a congressionally mandated comprehensive plan. Titles I and II might be kept at reduced levels, but the river basin commissions should expect to derive most of their funding from state sources. As for the new Office of National Water Policy, the Administration was determined to keep the concept of independent project review, but how and where had not been determined. The current chairman of the National Governors Association was to meet with Watt and key members of Congress shortly to protest these moves, but the representations were not expected to prevail.

With nothing to lose, NERBC vice-chairman Ronald Poltak wrote Secretary Watt on April 10, 1981 to protest the water resources budget deletions. These would effectively eliminate NERBC as a viable regional institution, he observed. In fact, state members were even more concerned about the loss of their Title III planning funds, a much valued source of help for state water resources investigations. Mr. Watt remained unmoved by these and other entreaties.

On the home front, the plight of regional agencies in general was of growing concern to the New England governors. Preserving NERCOM, with its annual budget of $7 million, most of it discretionary with the governors, was the priority order of business, but even this seemed doubtful. So useful had it become to the governors that they had allowed their own NEGC to lapse, taking regional action as necessary in their parallel capacities as state members of NERCOM. Serious discussions began about how to continue regional activities in the face of projected federal budget cuts. A New England governors research institute and a revived New England Governors Conference were suggested. Senior NERCOM staff were particularly active in these discussions since their own jobs were at stake.

On April 7, 1981, papers were filed establishing the New England Governors Conference as a Massachusetts not-for-profit corporation. Although the bulk of the purposes emphasized economic development, there were planks relating to planning, research, and coordination, and specific mention of natural resources and environment.

By late April, NERBC's Staff Transition Committee had devised a plan for continuing the commission through FY 1982 using state funds only. Approximately $200,000 in state contributions were projected for the regular NERBC program, and these would be sufficient for a modest staff of four professionals and operating support if commission and committee meetings were kept to a minimum. The March phase-out plan would have to be amended, however.

Alternate chairman Richmond felt obligated to begin sharing with the governors the explorations being undertaken by NERBC. In a letter dated April 27, he gave details of the situation in Washington and expressed some optimism. Three weeks later, he had to revise his earlier estimates sharply. NERBC is "virtually dead,"[84] he bluntly advised the governors. Secretary Watt wanted WRC to terminate the river basin commissions officially, and state members could not see their way clear to continuing state contributions in the face of their own severe budget cuts. Richmond could report, however, that the citizen advisors to NERBC were preparing recommendations for a continuing water resources function within the Governors Conference, possibly financed by a transfer of NERBC's remaining state and federal funds.

Commander Richmond's letter to the governors was personal and candid.

I have seen the free, easy and accurate flow of information between the several agencies and the several states to the benefit of both the individual and the group of states. I have seen the important water issues of New England discussed openly, frankly among those persons, federal and state, whose decisions and cooperation had far-reaching impact on the region as a whole and the states and specific areas of the states individually. Most importantly, I have seen the states submerge their parochial interest and prioritize those projects which are of greatest need to New England regardless of who the primary beneficiary may be. I have seen the priority problems of the area handled expeditiously to provide needed assistance to the states in a timely fashion; such as hydro siting, and drought management. I have seen the meticulous building of consensus necessary to develop regionwide policy.[85]

Gloom remained the order of the day at NERBC headquarters in Boston. In an internal memorandum entitled, "Down the River with NERBC," staff director Brown noted insult added to injury, for the House Appropriations Committee had recommended recisions in the FY 81 budget that would cost NERBC $160,000 of its current funds. Brown authorized the preparation of an option document of alternative institutional arrangements, which included NERCOM (if it was continued), the New England Governors Conference, the two existing interstate compact organizations (NEIWPCC and the authorized but inactive New England Interstate Planning Compact), the earlier federal-state compact (New England Water and Related Land Resource Compact) endorsed by three of the states and approved by the House but not the Senate, the Federal Regional Council (if continued), river basin organizations such as the three flood control compact commissions and watershed associations, and nonprofit organizations like the New England Natural Resources Center, the New England Rivers Center, and the New England Council. The option paper also summarized arguments for a continuing regional water resources program.

On June 4, 1981, NERBC's Steering Committee on Public Involvement, representing twenty-two prominent citizens from the six New England states, wrote the governors recommending a continued water resources coordination program within the New England Governors Conference financed by NERBC's remaining assets. In parallel action, the CCJP drafted a formal resolution for the governors to adopt at their June meeting. The federal members of NERBC met separately on June 18 to pledge their cooperation and full participation should the governors establish such a program.

On June 18, 1981 alternate chairman Richmond did, in fact, schedule a commission meeting—the fifty-eighth in its history and one that would prove to be its last. He reported no change in Washington. Secretary Watt was still prepared to ask for a vote of WRC terminating the river basin commissions. He advised that the NERBC Steering Committee on Public Involvement had urged the governors to consider incorporating water resource coordination under the wing of NEGC, and that a resolution to this effect was on the governors' June 25 agenda. State members had voted their concurrence and would carry out their responsibilities in continuing water resources planning among the states. The commission's Federal members had agreed in caucus to support the concept of a continuing water resources forum.

The commission then heard detailed reports on dredge management, ports and harbors, hydro power, water supply/conservation, and basin overviews. The Connecticut River Advisory Group hoped to have a final update of the 1980 Connecticut Basin Plan in the hands of the commission by September 1.

The next quarterly meeting of the commission was tentatively scheduled for September 24, 1981.

By July 1, alternate chairman Richmond could advise the NERBC family that the governors had adopted the resolution as requested—just in time, for on July 10 WRC completed official action on a memorandum of decision terminating the river basin commissions. NERBC would cease to exist on September 30, 1981 upon confirmation by presidential executive order.

At its July 29 meeting, the CCJP Committee approved a plan to dispose of the assets of NERBC. Its extensive technical library would be shipped to the Corps of Engineers in Waltham, Massachusetts, and the financial assets transferred to NEGC. Given approval by OMB, the disposition plan could be effected immediately.

On August 18, alternate chairman Richmond wrote each NERBC employee an official termination notice. He spoke of his regret at the decision to close down the commission and thanked the staff for their sustained and valuable efforts.

As expected, President Reagan signed Executive Order 12319 on September 9 terminating the river basin commissions.

On September 10, the CCJP Committee met again to finalize its program recommendations to the governors. The federal members were polled separately. Eight days later, the New England governors met and adopted a resolution approving the establishment of the New England–New York Water Council. It was careful to state that the council was solely advisory to the governors, and that staff would be appointed and report directly to NEGC. Any continuing commitment would be subject to the future availability of funds. The resolution specifically invited participation by the state of New York.

The governors chose to adopt a second resolution covering the programmatic details. The recommended program and budget for FY 82 were approved in the form drafted by the CCJP Committee. The chairman of the Governors Conference was authorized to enter into an agreement with NERBC for the transfer and use of its remaining funds. The current state members and alternates were asked to constitute the new council pending changes by the governors, and these members were instructed to prepare recommendations for staffing and other matters for "the orderly and timely initiation of the water resources program."

With termination just around the corner, alternate chairman Richmond could begin the final round of official actions. He wrote to his NERBC colleagues thanking them for their efforts and urging continued participation. He wrote to Secretary Watt as chairman of the WRC, ostensibly to give information about the infrastructure of New England water resources affairs, but also to convey the news of the new council and his continued personal conviction of the importance of planning. "I wish you God's counsel,"[86] he concluded.

On September 15, alternate chairman Richmond wrote NERBC members a brief memorandum describing the steps taken to achieve an orderly close-out of commission business.

For NERBC's staff, this was a difficult time. Many were young professionals deeply committed to the agency's work. It would be hard to find equally satisfying employment. A month later, twenty-one of the fifty employees were drawing unemployment compensation, unable to find the professional challenge they sought. The task of closing up an agency was onerous as well as distasteful. Records had to be sorted through, discarded, or packed. At the same time, the pressure was on to produce meaningful final reports on studies in progress.

Richmond's final communication was an open letter to citizens.[87] He ascribed NERBC's demise to its inability to compete in the federal budget. He spoke positively of the proposed New England–New York Water Council and hoped that it would become a "no nonsense Council that produces results." The citizens of New England and New York had the opportunity to enunciate their needs, he wrote, but, to be effective, citizens must demand and then support their new agency.

As the doors closed on the New England River Basins Commission and its two decades of history, the record fairly cried out for appraisal and evaluation. What was this era all about? Was it significant—useful? What does it say for bioregionalism in general and New England in particular? Will the past again be prologue to the future? The purpose of the remaining chapter of this report is to answer at least some of these questions. The water resources institutions themselves deserve analysis, and so do certain of the external factors that helped determine the shape and substance of the water programs during this period. Of these, special mention will be made of the New England governors, the water resources professionals, the water resources leadership, and the private sector.

The New England River Basins in Retrospect

THE WATER RESOURCES INSTITUTIONS

The New England–New York
Interagency Committee (NENYIAC)

Strengths and weaknesses underlie any venture of magnitude, and NENYIAC was certainly no exception. Despite an extraordinary effort at significant dollar cost over an extensive period of time, there is little to show for NENYIAC some twenty years later. To be sure, the forty-six volumes of inventory information are a monument in themselves, and participants are loyal almost to a man to NENYIAC's alleged virtues. Yet the study was, in many respects, essentially a bureaucratic exercise. Not even in its heyday was a serious attempt made to implement the findings. NENYIAC lacked political interest and support of any magnitude, and at no time did it appear to command a high priority with either the New England governors or the federal establishment.

Was NENYIAC then a mere venture in futility? The answer seems clearly in the negative, but for quite different reasons than one would suppose. New England, it is argued, received important and even vital benefits from the study, which greatly enhanced NERBC's chances of accomplishing its own objectives.

Before NENYIAC, the region could boast few effective working relationships in the water resources field either among the federal agencies themselves or between the states and the federal government. Participants freely admitted only a vague acquaintanceship with their counterparts in other agencies. Thus federal agency attitudes tended to reflect the biases of their national offices, unaffected by any personal working relationships within the region itself.

On their part, the state water resource agencies were just coming into their own in response to a series of natural disasters. The new interstate water pollution control compact and the regional investigations that followed periodic floods did encourage contacts between professionals, but only on a periodic basis. It would be safe to say that NENYIAC not only regularized these contacts but actually encouraged them by describing a need for a common front in the face of the projected federal onslaught.

It seems incongruous that an investment of the size of NENYIAC might be justified in substantial part by improved interagency relationships, but this is not far from the truth. For what NENYIAC did prove, echoing Leland Olds's earlier observation, was that federal-state cooperation *could* accomplish meaningful results even in a region as resistant to federal programs as New England. This heritage of close and effective working relationships was an asset of incalculable value as NERBC began its assigned mission.

A second significant contribution made by NENYIAC was the proof that coequality between federal and state representatives would work well in practice. In this sense, the New England venture appears to have set a useful precedent for the country as a whole. Coequality, of course, did not just happen. Neither was it the outgrowth of rational and deliberate action on the part of participants. The federal hand was, in large measure, forced by the tenor of the times. NENYIAC was in trouble at the outset, trouble that would clearly worsen if some concessions were not made to the state representations.

Moreover, the Corps of Engineers was being pressed nationally on the question of single agency leadership in river basin investigations. It had to be able to produce a creditable report in order to justify its claim to preeminence.

State objectives in seeking coequality were clear. With the chairman agency reserving its right to vote, the states could now force a standoff on the unwanted federal power and water development projects. Curiously, however, the record is quite different. Once the potential for negative action was assured, few of the executive council decisions were made by majority vote, most actions occurring by unanimous consent. And where matters were brought to an official vote, the pattern was invariably one of mixed federal-state balloting. On the larger scene, circumstantial evidence, if not actual proof, would indicate that the NENYIAC experience had a decided hand in the principle of coequality affirmed later in the Delaware River Basin Compact and the Water Resources Planning Act of 1965. New England's river basin struggles preceded each of these events, and such influential national advocates as the Council of State Governments and the Interstate Conference on Water Problems were fully cognizant of, and party to, New England's aspirations.

It also seems fair to conclude that the federal experience in NENYIAC and similar field agency committees served to set the stage for the later concessions that would make official federal-interstate agencies a reality.

Although hardly precedent-setting in its contents, the NENYIAC report must be regarded as a competent product and perhaps better than others of its kind. The inventory of resources was complete to the point of being voluminous, and, in some respects, its approaches can be regarded as precursors of the later comprehensive studies. Although often accused of "carrying water on both shoulders,"[88] the report did contain much valuable information never before compiled. More importantly, it offered instances where alternatives only were set forth leaving the actual implementing decision to a later political arena. This approach tended to reflect the growing body of opinion that would have river basin planners concerned primarily with the framing of alternatives.

In its approach to the sensitive problem of power development, NENYIAC also acquitted itself well. The compromise hammered out between public and private power interests, now somewhat academic because of the emerging nuclear technology, was a substantial achievement of its time and indicated that the give-and-take process could become politically viable even under difficult circumstances.

Despite its relatively limited distribution, the NENYIAC report has also proved useful in many subsequent studies, its base data serving as a springboard for future investigators.

Perhaps the least heralded of the NENYIAC virtues was its impact upon the states themselves. In most instances, state water resources functions were still in their infancy. The state administrators were surprisingly competent, such conditions notwithstanding, but they tended to lack visibility and stature within their own jurisdictions. Exposure to experienced federal personnel, plus the need to grapple with the broad dimensions of a regional study, left a clear mark on all participants.

Furthermore, New England had been split for years on a north-south basis, a heritage of efforts in the past to store water upstream for downstream beneficiaries. NENYIAC forced a careful look at regional water resources needs and a growing awareness of regional water resources problems. The emerging acquaintance with one another's problems, plus a need for a common cause against a federal enemy, forged a working relationship among state water resources administrators that was to bear continuing fruit in the years to come.

Finally, state and federal participants alike have spoken with great respect of what they learned from one another during NENYIAC. The dialogue between foresters, engineers, wildlife biologists, power specialists, and transportation and agricultural technicians could not help but rub off in significant amounts. For sheer breadth of exposure, NENYIAC was an enormously valuable experience for all participants.

The Northeastern Water and Related Land
Resources Compact (Northeastern Resources Compact)
The highlight of the NRC period was its pursuit of a water and related
land resources compact that would join the states and the federal government in a common cause. This has been represented as the first
serious effort in the nation to obtain a federal-interstate natural resources compact. From this experience, New England learned the
hard way about regional and national politics, but it lived to see its
recommendations reborn in federal legislation. A valuable and continuing alliance on water was formed in response to the compact proposal, including such diverse entities as the New England Council
and the League of Women Voters. Despite alternating hope and bitter
disappointment in the course of the four-year fight for compact ratification, the region's reflexes and perceptions were sharpened by ad-
versity.

One of the stranger paradoxes of politics must lie in the vigorous
sponsorship of the proposed northeastern compact by Rep. John W.
McCormack of Massachusetts, whose personal exposure to water resources was bounded by Boston harbor to the east, the Charles and
Muddy rivers to the north, and the Quabbin water supply aqueduct
to the west.

A South Boston state representative and senator, and (briefly) practicing attorney, McCormack began his rise to national prominence in
1926 by being appointed to the unexpired term of the late James A.
Gallivan of Boston, a district that he represented in Congress from
then on. For twenty years, he served as House majority leader,
House Democratic whip and, for six and one half years, as Speaker of
the House.

McCormack's intense concern for the northeastern resources compact can be explained neither by his personal interests nor by those of
the district he represented. Yet, without his personal determination
that the compact should see the light of day, the measure probably
would not have emerged from the Public Works and Judiciary committees and certainly would never have been approved by the House
as a whole.

McCormack's support for the compact was probably due to a combination of factors. He was, of course, urged to support it by business
and government leaders back home and thus, for normal political reasons, would have been inclined to lend the assistance of his office.
But McCormack also entertained an inveterate fondness for New England as a whole, often taking the leadership to advance programs in
the region's interest. His activity on behalf of fisheries and textiles, for
example, traditionally transcended the limits of his own district.

Further, a search of his legislative record reveals several significant actions with respect to water resources. For example, he entered the debate over the 1938 Connecticut River Valley Flood Control Compact by raising the first suggestion of a national flood control program unsupported by local contributions. McCormack was also prominent among the legislators who urged President Truman to establish NENYIAC by executive action, once the Senate had indicated that it would not accept the proposed Green amendment in the 1950 Flood Control Act.

Finally, there was apparently something in the entrenched opposition of the federal agencies to the northeastern resources compact that just plain stirred McCormack's Irish. He found the attitude of the agencies irrational and inconsistent and made this plain by deliberately pushing the compact proposal through the House.

For want of a better scapegoat, many NRC participants are apt to lay the full blame for the compact's demise at the doorstep of Sen. George D. Aiken, a scholarly-appearing, soft-spoken, self-styled farmer born and brought up in Vermont's scenic West River Valley. A wildflower grower by profession, Aiken cultivated an interest in politics early in his career, serving first in the Vermont House of Representatives (Speaker, 1933–35), later as lieutenant governor (1935–37) and governor (1937–41), and assuming a seat in the U.S. Senate in 1940 in fulfillment of an unexpired term.

In support of their conclusions, Aiken's critics assert that he lobbied most effectively against state ratification of the compact, that he organized a large part of the opposition testimony at the congressional hearings, that he worked against the measure within the New England congressional delegation, and that his influence as a ranking member of the Senate Agriculture and Forestry Committee helped persuade the Department of Agriculture to remain cool to the proposal.

One fact is clear, however. If Senator Aiken was not the compact's actual cause of death, his involvement was certainly a major complicating factor! The depth of his opposition can be well documented. His scathing testimony before the House Judiciary Subcommittee in September 1962 left no tone unstern. The sessions of the New England Senators Conference he attended were marked by substantial altercations over the compact issue. The numerous overtures made to him during the course of negotiations failed to reveal a single glimmer of compromise in either his public or his private posture. His official biography, distributed by his office, listed the defeat of the compact as one of the outstanding accomplishments of his career.

The real puzzle was not the Senator's opposition to the compact bill as much as the uncompromising nature of his attitude. This was made inexplicable by his later espousal of the Water Resources Planning Act, which set in motion the very coequal, federal-state body that he had opposed so vigorously. Yet, Senator Aiken was a respected and accomplished legislator, rarely known to be illogical in his actions.

One can try to attribute his adamant position to possible pressures from constituent groups, but the answer clearly does not lie here. Although the private utility leaders in Vermont were opposed to the compact, the senator had never been kindly disposed toward their interventions, being more closely allied with the electric cooperative and agricultural interests of his state than its business and industry groups.

Moreover, he belonged to that rare and vanishing breed of politician who enjoyed a virtual self-constituency. His campaign expenditures were nationally renowned for their thrift, and his reelection campaigns usually revealed a singular lack of substantive issues or opposition.

The real roots of Senator Aiken's opposition to the compact seem to lie in two areas: 1) his previous conditioning in federal-state water resources relations, gained during the early flood control battles over the West River in Vermont; and 2) his continuing distrust of southern New England business interests, as represented by organizations such as the New England Council.

One must recall that at the time of the Flood Control Act of 1938, Vermont was fighting federal flood control policy almost single-handedly. The Corps of Engineers had just won authorization from Congress for a series of flood control reservoirs in the West River Valley, and the new provisions of this act made it possible for the federal government to acquire and construct facilities even over the objections of the state in which the reservoirs would be located. The downstream states of Massachusetts and Connecticut, mindful of the damages suffered in the 1927 and 1936 floods, were anxious for the federal program to begin, viewing the federalization of the flood control efforts as the only way they could overcome northern New England opposition to the construction of the needed reservoirs.

By a curious twist of fate, the two Vermonters who would play the most prominent roles in winning modification of the 1938 Flood Control Act provisions were serving on the same state legislative committee in the early 1930s. State Sen. George D. Aiken of Putney and State Rep. Philip Shutler of Northfield sent to a swift death a bill that would have established a series of river administrative districts in Vermont, arguing successfully that government regulation of river development was not in the public interest.

There is reason to believe that Senator Aiken never quite forgot these bitter disputes over flood control, the early regional differences encountered in the drafting of a Connecticut River Valley Flood Control Compact, and the (at times) heavy-handed approaches of the federal agencies. His distrust of southern New England business interests had been fortified still further by disagreements over the St. Lawrence Seaway development and the determined resistance of the private utilities to public power.

Perhaps the support of the compact by Massachusetts Rep. John W. McCormack, a city-bred individual with reputed Boston business and utility connections, represented the final confirmation of his suspicions. At any rate, in an act of the utmost irony, Vermont's senior senator, architect and champion of most of New England's progressive interstate compacts, became the chief executioner of the proposed Northeastern Water and Related Land Resources Compact, a proposal later declared to be a pioneer in the field of federal-state relations.

Yet, besides Senator Aiken, there were clearly other factors that also militated against acceptance of the interstate compact. For example, two states (Maine and Vermont) never did ratify the compact, which raised uncomfortable questions as to the extent of New England's real interest in the proposal. Two other states failed to make any appropriations toward the proposed compact commission.

In contrast to the Delaware compact proposal, which saw the congressional delegations from four states and two major cities united firmly behind it, New England's political leadership also left much to be desired. It was apparent from the difficulties in Congress that the New England governors never really did make the compact an item of the highest regional priority.

Finally, the compact proponents lacked the greatest political action stimulant of all—good political timing. By 1961, Eisenhower conservatism had given away to the Kennedy era of experimental federalism, and the program emphasis had swung away from the states toward Washington. A bare half-decade later, however, water would again be in the spotlight, thanks to the unprecedented northeast drought, and New England would have the New England River Basins Commission, if not its long-sought Northeastern Water and Related Land Resources Compact.

NRC participants and observers are divided in their opinions as to the significance of the compact issue. Some regard the events as most unfortunate, stating that four years of valuable time were wasted and characterizing the New England determination as unbridled and unwarranted stubbornness. Others, however, regard the rejection of the northeastern compact, yet subsequent passage of the Water Resources Planning Act, as the height of inconsistency on the part of Congress and an uncalled-for slap in the face to regional interests. Still others have termed the compact proceedings and, in fact, NRC itself as a difficult but necessary period for the region to undergo in order to be ready for NERBC. The observation has also been made that NERBC was potentially superior to any interstate compact device advanced previously by the New England states.

The record would seem to indicate, however, that while valuable time was lost in the compact cause, New England's efforts helped bring about useful modifications in the Water Resources Planning Act. Although a comparison of the provisions now reveals few major differences with the earlier compact proposal, it must be remembered that the first versions of the act offered few concessions to state and regional interests.

In short, it would appear that NRC and its proposed interstate compact were far from lost causes.

The Northeastern Resources Committee (NRC)

Many of the criticisms addressed to NENYIAC would more than certainly apply to its successor, NRC. In its own declaration of intent, NRC was to bring about improved coordination, resolve conflicts between agencies, adjust conflicts in interests, and promote state and federal programs and policies in accord with regional needs. Virtually none of these worthy objectives was accomplished to any appreciable extent. Lacking a central staff and budget, NRC could not even get its own house fully in order.

Its larger objectives—to develop a progressive program that could and would be supported by the region—never materialized at all. As one observer put it, "If NRC accomplished much of anything I do not know what it was."[89] Others, however, have responded differently— one in fact labeling NRC "an exceptionally valuable instrument."[90] Among its accomplishments, the following can perhaps be singled out for special mention.

In contrast to NENYIAC, NRC was sought after and actually fought for by its participants. Its foundations were thus solid and credible. Although part of the incentive was clearly negative—anxiety that a less self-determined vehicle would be imposed upon the region—the predecessor, NENYIAC, had done its missionary work well. Even the most vested of earlier opponents had come to recognize the need for coordinated water resources planning.

NRC's intentions were also clearly well founded despite an occasional lack of incipient goals and missionary zeal. Though it was often faulted for not being sufficiently aggressive, its series of state meetings, for example, was an impressive venture for an organization lacking central staff or funds.

NRC, as it were, "kept the store open"[91] during the interim period, maintaining valuable lines of communication between the states and the federal agencies. As new personnel came along, the machinery was at hand for proper indoctrination and continued coordination.

Once the compact issue had been settled and a suitable alternative was available, NRC had both the experience and the heritage of working relationships to develop a practical approach for the New England River Basins Commission. In retrospect, New England was probably readier for the new vehicle than any other region in the country—or even the national administration in Washington.

The New England River Basins Commission

The commission offered opportunities for maximum participation by several levels of government, thereby bringing the utmost sensitivity and timeliness to water resources planning. Its provisions combined the tested principle of coequality with a framework of permanence NENYIAC and NRC never could attain. In many respects, it was the regional water and related land resources agency, now enabled nationwide, that New England had never been allowed to institute for itself.

Regional coordination of interests and efforts, in turn, was expected to produce projects that would work and would be acceptable. Acting collectively, the region could begin to exert a block influence on the program authorizing process, subverting individual agency influences and wielding the most effective legislative weapon of all, the interest of the folks back home. The alternative seemed to be continued ineffectual national water programs, executive agency empire building, senseless quarreling within the region on water matters, and single-minded control by Congress.

For New England, wedded firmly to its New England River Basins Commission, the job was to make the new machinery work. The commission, of course, must remain credible—from the chairman himself to the professionally high standards demanded from his staff. As the federal representative on the commission, the chairman must also be able to bring a firm federal commitment to the conference table, either through his personal influence or by the backing he was given by Washington. In turn, the various state delegations must bring from their respective governors and state legislatures a clear commitment to support the work of the commission. The state representatives must also have sufficient stature and professional capabilities to deal effectively with lesser jurisdictions of government and thereby to contribute meaningfully to the regional planning and policy programs.

NERBC's mandate by statute was, of course, quite broad, ranging from the preparation and maintenance of a master plan to the conduct of annual project programming. The experience of the past would indicate that the broader overview function would be the more productive of the two, with detailed planning and project action left to federal and state agencies. By stressing its review authority and developing, as it were, a virtual veto power over project implementation, the commission could become the single most powerful influence of all in relation to regional water resources programs.

A key factor in determining the success of the commission was certain to be the policies of the Water Resources Council (WRC) in Washington. The extent to which it tied the commission's hands by rule and regulation and, conversely, the willingness of the chairman to exercise his statutory right to go directly to the President, might well determine the value of the institution in the long run.

Judging from the past, however, it would also be wise for the commission to begin building for itself a secure public and legislative constituency, enabling it to move effectively and independently on the regional and national scenes. This suggested the desirability of a counterpart private organization, responsible enough to put the commission's programs to the test, and influential enough to see them through.

Participants and observers have listed a number of strengths in NERBC's subsequent approach and program. For example, the agency has received generally high marks for its information gathering, dissemination, and transfer functions. Deemed particularly useful were those relating to the collection and analysis of large-scale data, a necessary prerequisite for the assessment of regional impacts and the determination of regional futures. In fact, issue identification in a holistic framework would have been difficult to achieve otherwise through the conventional apparatus of state and federal agencies. The process of general consciousness awakening within the region was probably one of the major contributions of NERBC over its fourteen-year history.

Closely related to the information phases was NERBC's service as a forum for state, federal, and public participants. It had the capacity to pull people together to discuss mutual needs and concerns. It was a place where an outsider could usually find out what was going on. It was also a place where insiders at high policy or operating levels could share experiences and simply get to know each other better. For the regional personnel of the federal agencies, trapped within bureaucratic hierarchies, this kind of nonthreatening intercommunication was especially valuable. It provided the setting for a sense of reasonableness that never would have resulted from interagency agreements or statutory program directives. As one veteran federal observer put it, the likely enduring benefit of NERBC is apt to be simply its contributions to the cooperation of man.

A third set of strengths fell within the area of conflict resolution. On the rare occasions when it was invited into controversy, such as the matter of disposal of dredged spoil in Long Island Sound, NERBC acquitted itself well. Yet there were countless other occasions, much less formal in nature, when its functions in information analysis and transfer, and its availability as a forum, tended to quiet problems and keep them contained. It is hard to say whether conflict resolution was a deliberate or an inadvertent role for NERBC. On the one hand, there was the statutory requirement of consensus imposed upon it. On the other was the political reality that unresolved conflict could easily destroy the fragile balance essential to the functioning of such a regional body.

NERBC also performed valuable service for the region with respect to federal programs and policies. States could utilize it as an additional pressure point on the federal system, exerting leverage on funds and policy positions in ways no single state or organization of states could ever hope to achieve. But the reverse was also true. The federal agencies found NERBC useful as a means of reaching the states to gain support for programs, projects, or policies of particular interest to them.

The commission's staff and its program products were generally well received within the region. NERBC personnel were a cut different than their state and federal counterparts—young, bright professionals with a fresh coat of paint and a largely optimistic view of the future. Their enthusiasm for planning was unrestrained, their land use concerns broad and eclectic. Yet, as many of the program materials indicated, the staff could also turn out competent technical products. The commission's chief programmatic virtue was its flexibility. It was usually ready and willing to be of assistance when needed.

Yet there were weaknesses in the NERBC program, some clearly defined and others subject to interpretation. For example, most observers found it complex and hard to understand. It had no apparent direction or sense of legitimacy. The result was the absence of a meaningful role, or one that never got across to others. This led to a growing sense at the end of an organization that was spending time and money merely seeking things to do.

Closely allied to the problems of role and function was the absence of a legitimizing constituency. The commission was a mixed bag—a creature of both the states and the federal government. It never developed an effective public outreach of its own and, at the end, was bereft of support from any quarter. To the states and the few informed members of the public, it was primarily a federal agency. To the federal government, its alleged predilection for the states deprived it of a meaningful role in the struggles over national programs and policies. As a body without a defined power base in the federal system, the commission simply fell between the cracks.

The absence of authority, ensured by the consensus in its language in its enabling statute, virtually guaranteed that the commission would be viewed with indifference and even disdain. Lacking the power to do things to or for others, it was unable to bring even its own plans to fruition. The inability to implement led to its reputation as a paper tiger. It found itself unable to reach the decision makers in Washington (the Office of Management and Budget, the agencies represented on WRC, and the Congress), or the decision makers within the region (the cabinet-level agency heads and the governors).

Procedurally, the NERBC program was also flawed. Its direct control of funds was limited. The superstructure of committees, task forces, and work groups was cumbersome and unwieldy. Its meetings were frequently crowded with routine business and dull. And in a genuine effort to be democratic, it opened the door wide to so many interests that the net result was occasional anarchy, near paralysis, and an ultimate level of agreement often too modest to be meaningful.

On the question of avoidance of controversy, opinion is divided. The commission is accused of tiptoeing through the region, of covering up issues that could be controversial, and thereby suffering a loss of vitality of mission. But the record would indicate that it was rarely invited into sensitive issues, a case in point being the Dickey-Lincoln project, a regional issue that Maine steadfastly regarded as its own. By the time NERBC had become securely established, the practice of avoiding controversy had become a habit.

Similarly mixed are the comments on the role of the states. As the initial curiosity dwindled, the states seemed to lose interest and pass the mantle of representation to succeedingly lower levels of the bureaucracy. The attendance records verify this. Yet it must be remembered that NERBC's primary business was planning, that planners rather than line water resources personnel had the ear of the governors during much of the period, and that the vast majority of the business of the commission was appropriately the province of alternates rather than principals. Still others hold that the states were simply overwhelmed by the federal presence and were cowed by the handful of individuals who dominated the program. For those familiar with New England's famed Yankee independence, intimidation to such a degree is simply inconceivable. However, the states did ultimately decide that it was not worth investing the energy needed to redirect the commission's course.

In assessing what went right and wrong with NERBC, the passage of time has to have been a significant factor. The commission came into being in 1967 during the full flush of the environmental movement. Comprehensive planning was again in vogue, and water resources planning was an item of priority attention in the aftermath of drought and floods. By the early 1970s, the public's attention had been diverted to matters of environmental protection. EPA had come into being and, along with it, the staggering national clean water and clean air missions, responsibilities that it chose not to share with the river basin commissions. It was also a time for the oceans and shoreline — ostensibly a "related land resource" in the meaning of the Water Resources Planning Act but, in fact, an issue with its own political vitality and constituency. By the late 1970s, the region and, indeed, the nation were swept up in an energy crisis, and the spotlight was on the potential use and development of the Outer Continental Shelf. To NERBC's eternal credit, its staff and leadership recognized the need for program redirection in order to remain relevant; but in so doing, it harvested further ill will for engaging in what others regarded as peripheral matters and diversions from its primary water mission.

One program participant could have helped measurably throughout, and that was the federal WRC, the surrogate parent of all the river basin commissions. While WRC did not impede the progress of the commissions, it gave them no guidance either. In fact, some observers flatly attribute the demise of the Title II river basin commissions to the failure of the Title I council. Yet the record is mixed on this point too. Early in the water resources planning program, WRC developed draft regulations governing the activities of the Title II commissions. These were bitterly opposed by the chairmen as an unwarranted intrusion. Already-strained relationships were exacerbated by the fact that the chairmen held presidential appointments, while the executive director of the council did not. The net result was a program processing and budgetary relationship only, and a council staff increasingly preoccupied with the federal side of its mission, notably the coordination of federal programs, the setting of federal project priorities, and the development of a national water policy. The failure of WRC in the national arena further compounded its difficulties with the river basin commissions. If the council exercised no clout with the federal agencies, OMB, and the Congress, how could it be expected to help commissions like NERBC? Lacking a definitive set of expectations either by statute or regulation, without provisions for oversight or performance audit, and with no one to report to, NERBC found itself increasingly isolated.

The obvious solution was to turn to the governors of the states for guidance. Indeed, some observers have characterized NERBC as virtually pandering to the states to the detriment of its federal responsibilities. In attempting to reach the governors and earn their confidence, NERBC struck out again. The New England governors' view of the commission, fueled by the growing dissatisfaction of its state members, was that of a fuzzy regional study group at best, and an ineffectual body at worst. Unlike the New England Regional Commission, which had supplemental federal funds to distribute within the region, NERBC could do little to help the governors. It was, therefore, difficult even to get their attention. This meant that NERBC was increasingly disenfranchised — ineffective as an instrument for reaching decision makers at either state or federal levels. In an environment where the exercise of power is the name of the game, at the end of its tenure NERBC was hardly even a player.

In Summary

The institutional structures attempted in New England over the thirty-year span of water resources history have been varied in design and execution. The first (NENYIAC) was purely federal but modified in practice to provide for state representation. The second (NRC) had dual credentials from the Interagency Committee on Water Resources (IACWR) and the New England governors but lacked legislative sanction. The third (the proposed northeastern resources compact) would have enjoyed formal status as a federal-interstate agency. The fourth (NERBC) took advantage of federal enabling legislation to create a federal-state entity. The fifth (The New England–New York Water Council) is a state institution with allowance for federal participation. As yet, none has worked satisfactorily. The process of experimentation continues. The simple truth appears to be that a fixed institution, without the capacity to adapt itself to changing circumstances, is destined for eventual obsolescence.

THE NEW ENGLAND GOVERNORS

In reviewing the actions of the New England governors during the NENYIAC, NRC, and NERBC years, four major periods of involvement can be identified: the early stages of NENYIAC, the interim period between NENYIAC and NRC, the occasion of the proposed water and related land resources compact and, finally, the operation of NERBC. In each of these periods the governors were required, first, to determine specific regional policy and, second, to carry out a tangible course of action.

With regard to the first period, the records of the New England Governors Conference (NEGC) give no indication that the governors were approached as a body on the various proposals for valley authorities and survey commissions prevalent during the late 1940s. In fact, the adverse reaction of the governors following the establishment of NENYIAC would indicate that most (with the possible exception of Democratic Gov. Paul A. Dever of Massachusetts) would not have favored such an approach if they had been asked. In the light of the apparent lack of support from within the region, why did so many individual members of the New England congressional delegation pursue such a course? The answer seems to lie in a number of related areas.

Firstly, the economic recession of 1949–50 had made the region's political leaders development-oriented. The governors, for example, had launched an ill-fated attempt to obtain a New England Development Authority by interstate compact action. The failure of the compact in the state legislatures left the federal government as the next most likely place to seek the desired economic planning and development assistance. Secondly, the New Deal influence, though in its waning days, still cast a substantial aura on the Washington scene. Water resources development legislation was in vogue, and the New Englanders could hardly fail to participate in some fashion in this national movement.

Thirdly, despite the general skepticism prevailing, there was appreciable sentiment back home for regional development of water resources. Labor groups, for example, made strong representations in this regard at the urging of their national headquarters. The New England affiliates of organizations such as Americans for Democratic Action and Judson King's National Populist Party also produced a steady drumfire of propaganda for such approaches.

Lastly, there was really no concerted regional leadership in New England at the time. Linkage between the congressional delegations and the governors had traditionally been on an ad hoc, state-by-state basis. The machinery was just not properly developed either to produce an expression of consensus, or to convey such results forcefully to the region's congressional spokesmen. In some instances, the governors were considered potential political rivals.

Once the threat to New England's autonomy had been put to rest, the governors seemed to lose interest in NENYIAC. For example, in sharp contrast to the earlier round of public hearings, only two governors, those from Maine and New Hampshire, attended the final round of hearings in which the NENYIAC findings were made public, and only then to present formal and perfunctory greetings to the groups assembled. Although each of the state representatives attempted to keep his chief executive fully informed, the state members came to speak more for themselves than for the states they allegedly represented.

Furthermore, what little the governors did know of NENYIAC related principally to individual state problems. Although on at least two occasions briefing sessions were arranged for the governors by NENYIAC officials, there was no continuing mechanism to keep the regional aspects of the study before NEGC. Toward the close of NENYIAC, however, it became rapidly evident that communications would have to be improved. The provisions of the 1944 Flood Control Act required that each governor submit official comments on the final document.

Lulled by the generally encouraging reports received from their individual designees, the governors tended to regard NENYIAC as of rather low priority. The state representatives, in turn, having brought about what they regarded as a remarkably restrained document, furthered this concept by submitting bland and generally complimentary drafts of official comments for the governors' signatures.

Advised by their official representatives that NENYIAC was to be an inventory only, the governors were disinclined to press for any implementing action. It seemed prudent to leave well enough alone. Continued federal-state coordination was recommended by some of the governors, but as much to keep the federal agencies in full sight as to advance the joint planning approaches made effective by NENYIAC.

Perhaps the most striking evidence of the prevailing attitude came with the 1955 floods. Given a period of proper preconditioning by NENYIAC, plus a strong collective interest in water resources development on the part of the governors, the floods might have been expected to serve as the ideal springboard for implementation of NENYIAC-identified projects. Such was not to be the case. While the New England governors did present a strong and united effort to Congress for flood relief funds, supported with considerable unanimity by the New England congressional delegation, the avenues employed bore little direct relationship to NENYIAC. Although much of the previously gathered data undoubtedly proved useful, it was the New England Division of the Corps that supplied the actual project information. No serious consideration appears to have been given to reconstituting NENYIAC for even a short period as a coordinating flood relief agency.

During the interim period when state representatives were seeking a suitable successor to NENYIAC, the governors' interest seems to have quickened again. Much of the stimulus came from Gov. Dennis J. Roberts of Rhode Island, who was then serving as chairman of NEGC. A practicing attorney and former mayor of Providence, he had little in his background to suggest any major interest in water resources. However, his leadership position during the 1955 floods, his increasing knowledge of NENYIAC through contacts with the state members and the Permanent Regional Committee reviewing the study, and his subsequent contacts with the Washington-based IACWR, all led to a growing appreciation of what was at stake.

The New England governors, as before, had great difficulty in formulating a forthright regional position. While all were in favor of some vehicle for continued coordination, many were reluctant to embrace the militant position of the state representatives, particularly after a delegation from IACWR had spoken convincingly of the ease with which the new regional organization could be chartered by Washington. However, IACWR could not give the governors a clear commitment on the federal policy position until the Presidential Advisory Committee on Water Resources Policy (the McKay Committee) had rendered its final report.

Once it was clear that the Administration would embrace a program of stronger federal-state river basin committees, Governor Roberts lost no time in requesting a Northeastern resources council on behalf of NEGC. IACWR responded in short order by approving the issuance of a formal charter, and the Northeastern Resources Committee was in being.

Despite the relatively extensive correspondence and discussions on this issue at the time, subsequent events revealed that few of the governors were fully cognizant of what their actions really meant. NRC's quest for a written expression of approval from the governors brought this situation to light a few months later. More than two years elapsed following the issuance of the charter by IACWR before the governors could be persuaded to enter into a formal memorandum of understanding regarding NRC. Furthermore, when finally approved, the language of the document was studiously careful to circumscribe the extent of the commitment by the New England states.

During the course of NRC, periodic efforts were made to apprise the governors individually of progress being made. The round of state public meetings held in 1958, most of which were attended by the governors, afforded tangible evidence of program activity, but it was not until the state members decided to seek a formal interstate water and related land resources compact that the governors were approached officially again. Preconditioned by a previous heritage of compact action and briefed in advance by various NRC state representatives, the governors offered no objections to the proposed compact, even though it reinstated the individual state financial commitments they had so carefully eliminated in the NRC memorandum of understanding a year previously.

Once again, the actual depth of understanding and support on the governors' part proved illusory, for the NRC members encountered substantial difficulty in getting the individual enabling acts even introduced into the state legislatures, let alone enacted. When opposition began to develop in Congress toward ratification, the lack of real commitment and interest by the governors became painfully apparent. Had it not been for the ratification actions of four state legislatures, which virtually required official endorsement of the compact by those chief executives, it is doubtful whether NEGC could have been prevailed upon to support the compact at all.

By the time of the Water Resources Planning Act and NRC's decision to endorse this route for its river basin planning efforts, the governors were in a receptive frame of mind to approve the new approach. There were several reasons for the marked change in attitude.

In the first place, little real support remained for the interstate compact and, in fact, there was considerable sentiment within NEGC for a thorough reorganization of the compact organizations already established within the region. Gov. Philip Hoff of Vermont was particularly vocal in this regard, as the governor with the most direct interest and the most direct experience in water resources problems. Furthermore, the region was entering its fifth successive year of water shortage. Espousal of any water cause at this point was politically attractive.

Thus a combination of skillful selling by NRC proponents and the timeliness of the proposal itself brought about genuine support for the projected New England River Basins Commission from the governors. The unexplained delays on the federal side, if anything, only whetted their determination to see it established.

Having brought NERBC into being, the governors lost little time in moving on to other projects. Their real attention was focused on the New England Regional Commission, also established at their request, in 1967. Here their involvement was direct and real, for under the provisions of Title V of the Public Works and Economic Development Act of 1965, they were themselves the state members. With approximately $5 million in project funds to allocate annually, the governors could directly affect the course of events in the region.

Nevertheless, the record of the early years of NERBC indicate that the commission enjoyed the confidence, if not the constant attention, of the governors. Although the governors did not fully understand the program, the professional reputation of the chairmen was such that the requested support was usually forthcoming. It was customary to meet with the governors at least once a year, and it was the early practice of the commission to bring its program and budget to them for approval. In those days, chairman Frank Gregg felt confident that he could meet with any governor as needed. In fact, his personal relationship with a number of governors served to strengthen the influence of the official member from those states.

The chairman of the Governors Conference at the time of the commission's activation, Vermont's Governor Hoff, actually initiated a central theme that carried throughout the entire NERBC period. The governors were understandably exercised over the proliferation of organizations and activities at the regional level. Upon several occasions, they asked the executive director of NEGC, in conjunction with the chairmen of the two regional agencies, to explore ways of consolidating existing organizations. In later years, their concern about uncontrolled policy or program activity led to the actual withholding of state appropriations.

Upon rare occasions, individual governors would turn to the commission for assistance. The siting of major energy facilities, such as the Maine Yankee and Seabrook nuclear plants, warranted independent evaluation, as did the proposed pumped storage project in western Connecticut and Massachusetts. The governors asked NERBC to prepare regional recommendations for pesticide usage. Yet there were also occasions when the commission's prospective involvement in regional issues drew gubernatorial opposition — the Dickey-Lincoln project in Maine, and the exploration for oil and gas on the Outer Continental Shelf. For the most part, the governors preferred the New England Regional Commission as the source of fire-fighting assistance, for the NERCOM alternates were members of their personal staffs. But when the governors did speak, the NERBC members, state and federal, listened. It was a fact of life that state support was a crucial element of the commission's program.

NERBC's mixed record with the New England governors is more circumstantial than intentional. In terms of a constructive relationship, both institutions were flawed. As one governor has observed, the governors' accomplishments collectively have not been noteworthy. They tend to accept actions regionally only when they do not conflict with their political concerns locally. As for NERBC, lacking both program resources and clout, it could do little to help the governors. After a brief "Camelot" at the beginning when the states were represented by elected officials with a high degree of mutuality, regionalism fell out of favor with the governors. Under such circumstances, the only recourse for the leadership of the commission was to work through its appointed members. And as those representatives derived from successively lower levels of the bureaucracy, NERBC found itself increasingly removed from the attention of the governors.

At the height of its planning activity, the commission did have a momentary reprieve. State representatives had become either planning officials or the new generation of cabinet-level environmental officers, individuals who had potential direct access to the governors. Moreover, the large Level B studies commanded many millions of dollars, and a measure of influence over the federal resource agencies. During this period, planning was in vogue, and NERBC, as the region's chief planner, was singularly visible. Yet, lacking the power to implement those plans either directly or indirectly, it could not make its influence stick.

For much of its life, NERBC found the New England Regional Commission interposed between it and the governors. Despite countless memoranda of understanding, the sharing of office space, and even a period of joint vice-chairmanship, the two were never able to work well together. Each regional agency was chaired by an experienced, aggressive, federal official. Their missions were potentially in conflict — economic development for NERCOM and conservation for NERBC. Upon the rare occasion of a project in common, such as the siting of bulk power facilities, the transfer of funds to accomplish the study was blocked by congressional opposition. And the character of the two agencies was entirely different. NERBC was staffed by resource professionals with a planner's view of the future. By way of contrast, NERCOM's time frame was in terms of political immediacy. Its efforts at coordination were designed mostly to keep NERBC at bay. Small wonder then that the governors' impression of the regional agencies was one of constant staff infighting and jealousies. In many respects, this was true.

At the end, the New England governors did step in and ensure a continuing, regional, water presence. Three resolutions were adopted in 1981 establishing the successor New England–New York Water Council, but the governors were careful to make the program and staff directly subservient to NEGC. Had it not been for direct appeals to the governors from acting chairman Stephen Richmond, and intervention by the commission's citizen Steering Committee, it is unlikely that even this action would have taken place. What seemed to turn the tide was the availability of nearly $300,000 in unexpended state and federal funds. Even so, the governors were careful to include in their resolution no permanent commitment to the new council.

In Summary

In reviewing the record of the governors' participation in New England river basin affairs, one is tempted to write off their contributions as being generally ineffectual, placing the blame on an absence of real interest on their part in natural resources affairs. This is not an accurate appraisal, however. If a lack of affirmative interest in water and related land resources is apparent in the early record, it merely reflects the tone of the times. As political representatives of their respective states, the governors only mirrored a prevailing public opinion that seemed not fully convinced of the importance of natural resources and conservation programs.

By the time of NERBC, however, the public outdoor recreation and conservation ground swell had reached substantial proportions in New England, and the climate was decidedly ripe for efforts in this field. Although only a few of the governors took a substantial personal interest in water resources during this period, the caliber of the governors as a whole was extraordinarily high. Moreover, they came to work together remarkably well in the interest of the New England region, considering differences in party affiliation and sectional representation.

If the governors displayed any substantial failing, it lay in the nature of their organization, NEGC, which professed far more than it was able to practice. Meetings of the conference were often unsatisfactory—long on special interest items and short on advance preparation and time. It frequently served as a sounding board for individual governors or groups that they represented. In short, the conference was perhaps more used than useful!

Under these circumstances, it is a remarkable tribute to the governors, and those who worked to make NEGC an effective instrument, that the record displays as much action and unanimity as it does on regional water resources problems.

THE WATER RESOURCES PROFESSIONALS

Of all the meanders that marked the NENYIAC, NRC, and NERBC courses, none was more dramatic than the change in attitude displayed by the state water resources professionals as the regional program progressed. From outspoken opponents of federal intervention in any form, the state representatives changed course to become vigorous proponents of state-federal river basin planning, and later lost interest in what they themselves had helped create. The reasons for this change of heart represent an important part of the New England river basins story.

At the time of NENYIAC, New England's field of vision was narrow indeed. The experience of its officials was limited at best. Relatively new to the job of water planning, they tended to seek the more familiar ground of the status quo, responding to a conservative constituency comprised principally of industry spokesmen.

Furthermore, the traditional New England suspicion of central government in any form was not far from the surface. The past confrontations with the federal government, notably the hard-fought battles over flood control policy on the Connecticut River, had done much to sustain these traditional inhibitions. The federal assertion of power in the Flood Control Act of 1936, ameliorated though it was by the later state consultation provision of the 1944 act, was not to be forgotten for some time. The fires of independence were also stoked by a sincere conviction that New England could really do the job itself if left to its own devices. Relationships with federal personnel were only spasmodic at the time, hence hardly conducive to building confidence in interagency actions.

But perhaps the greatest instrument for change was the simple matter of exposure to one another over time. This process had been abetted by the all-important factor of continuity of personnel. Positive personal relationships developed during the NENYIAC and NRC periods that smoothed over rough edges and helped turn the federal involvement into an asset rather than a liability. As each individual action attained credibility, so also did the undertaking for which the group was collectively responsible.

Perhaps the key decision was made by Colonel Talley when he granted the state representatives a full voice at the NENYIAC conference table. To do so, of course, was not only a violation of federal precedent but also a direct denial of the role of observer that President Truman had outlined for the governors' designees in his letter of October 9, 1950. The state members were first suspicious, then astonished, and later enormously proud of the democratic manner in which the NENYIAC issues were debated and decisions arrived at. By giving the states their way, Colonel Talley had, as it were, built them into the NENYIAC effort. Once party to a decision, the representatives were obligated to defend the position even to adverse constituencies within their own states. This they did with a fervor that became a self-generating force for further consensus. By the time NENYIAC had come to an end, the state representatives literally could not conceive of any better way to conduct regional water resources planning than through a joint federal-state agency.

The magnitude of change was well illustrated by the NRC discussions over the proposed water and related land resources compact. IACWR, in defense of the federal position, suggested the alternative of a simple compact between the states without formal federal representation. Incongruous though their position would have seemed a scant five years earlier, the state representatives actually held out for full federal participation. The region would now have none of the federal government *not* becoming a full partner in planning and policy affairs.

By the time of NERBC, the state members were potentially more influential than the federal representatives. The history of previous efforts was well known, and several members of the new commission brought with them long institutional memories. Despite this auspicious beginning, the states were never able to respond to the opportunity at hand. They failed to manipulate the large, comprehensive studies in their own interests, and they rarely used their position on the commission as leverage with the federal agencies. In many respects, the bottom line of any appraisal of NERBC is less the failure of the commission than the failure of the states to take full advantage of it. There are many reasons why.

In the first place, the top echelon of the states was never active in NERBC's program. Since it was well run, it did not demand their attention. This problem was compounded by generally poor communications between the state representatives who did participate and the top levels in their own states. At times, the federal members and staff wondered whether the representation was more personal than official. To be fair, the states often felt like poor cousins, unable to share in the lavish planning studies because of constraints in the federal funding provisions. State budget cutbacks also limited their participation. In some states (e.g., New Hampshire), a hostile gubernatorial climate for much of the period made state representation no easy task. The absence of state plans that could serve as building blocks for regional studies left the states generally without a frame of reference for the large, Level B investigations.

Other factors contributed to a sense on the states' part that they were more tolerated than served. The rapid turnover of state personnel meant that there was often little continuity in representation. The final annual report (1981), for example, lists nearly eighty individuals who served at one time or another as state representatives or alternates – nineteen from Connecticut alone. And, as one top federal official observed, the states were curiously uninformed about federal procedures and functions. This put them at a distinct disadvantage in the setting of the commission.

The net result was that the states simply dismissed NERBC after the initial curiosity dwindled. It was viewed as a luxury at the end – a costly device with nothing to show in the way of state benefits. The simple truth seems to be that the states were never really sold on the need for a federally authorized river basins commission and, consequently, had no sense of ownership in it. Of those state members who did see the potential, many concluded that it was not worth the investment of time and energy to deflect the institution to a more useful role. Thus an interesting question could be raised: Are the states a natural constituency for a river basins commission established under national enabling legislation? In New England's case, perhaps the more modest, home-grown compact would have fared better.

With six interstate compact agencies in existence for more than a decade, one would have thought the region thoroughly preconditioned for a federal-state agency like the New England River Basins Commission. Yet, the record of the compact agencies under NERBC is a disappointing one. After the first flush of curiosity passed, few even bothered to send official representatives to commission meetings. When they did attend, the interstate members found little of interest to them except when their "turf" was threatened.

One of the earliest actions of NERBC was to try to define the rather fuzzy role prescribed for the federally sanctioned compact agencies by the Water Resources Planning Act. It was determined that they were empowered to vote on all matters except the election of a state vice-chairman. In a report to the commission during its first year, a committee of compact agency representatives reached the self-serving conclusion that theirs was so specialized a set of functions that it should not be disturbed by efforts at coordination.

On NERBC's part, no real incentives were at hand to disturb the status quo. Once chairman Gregg's move for a water quality committee had been rejected by a coalition of state, interstate, and federal members, it became more trouble than it was worth to involve the compact agencies meaningfully in the commission's program.

The role of the federal members of the commission also merits some attention. At peak, ten of the twenty-three official representatives came from federal agencies. Nearly one hundred federal officers served as members or alternates during the life of the commission.

During the course of debate over the Water Resources Planning Act, much was made of the potential threat posed by the federal presence. The coequal voting provisions and the consensus requirement were designed to help prevent federal domination. Yet, in the course of NERBC's entire fourteen-year history, there was never an instance in which the chairman and vice-chairman formally set forth their separate views in accord with the provisions of the act. There were occasions in which an agency exercised its prerogative of abstaining from voting, but, invariably, the voting blocs on issues consisted of mixtures of state and federal representatives. The other potentially worrisome federal dominance issue, the actions of the presidentially appointed chairman, never materialized at all. In fact, both full-time chairmen were viewed by the federal establishment as favoring the states.

Yet the federal presence could not help but influence the course of events. It was the source of the bulk of the commission's funds, state contributions notwithstanding. Because of this, program activities tended to follow federal procedures. The commission was also a creature of the Water Resources Council, a federal interagency body, and it performed regional assessments for WRC as part of its national responsibilities. The federal presence could also be a complicating factor in the case of sensitive issues such as Outer Continental Shelf development, for the agency chairing WRC (Interior) was the primary federal advocate for rapid exploitation.

In the end, though, it was the subtleties, rather than the realities, that prevailed. Despite a record of nonintervention by the federal agencies, and even indifference by some of the most powerful (EPA), the commission remained haunted by a reputation of being primarily a federal agency. That perception was shared by the states and the public to the very end.

In Summary

One cannot help but admire the tenacity with which the state representatives held to their convictions in the early years, seemingly hopeless odds notwithstanding. On the whole, history appears to have proved them right. Yet although the state representatives should be credited for their constancy, they have to be faulted for lacking the right political credentials really to put their convictions across. In terms of tangible output, perhaps the wrong people were making the decisions for New England much of the time.

Generally high marks must be given the federal agency personnel in New England during the period in question. Within the policy limitations of their respective agencies, responsible and sincere efforts were expended to make the New England operations a success. In some instances, for example, the drafting of the interstate compact, the degree of cooperation appeared to go well beyond the realm of normal duty.

New England also seems to have escaped almost unscathed the bitter interagency rivalries that characterized basin planning operations elsewhere. This is attributed, in large part, to the personal qualities of the officials assigned to the region, to the long history of cooperative endeavors, and to the continuity of relations brought about by NENYIAC, NRC, and NERBC.

This general heritage of accord also produced a curious esprit de corps among seasoned participants. Fences tended to be mended promptly when disagreements broke out, with federal and state representatives lending a hand indiscriminately in the common cause. In the case of the final NENYIAC report, there was a very real sense of pride and accomplishment in the final results. And even post-NERBC, federal and state members alike genuinely regret the absence of an agency that, at the very least, encouraged cooperative action by the region's water resources professionals.

THE WATER RESOURCES LEADERSHIP
Among the regional water resources officials who occupied a leadership position in New England at the time of NENYIAC and who were to feature prominently later on, the following are deserving of special mention:

Walter H. White, Chairman of the New Hampshire Water Resources Board, was by background a highly improbable spokesman for New England water resources interests. Registrar of probate for Carroll County, town meeting moderator of the village of Ossipee, and a long-time Republican county chairman, White arrived at his resources post through a series of political appointments, including commissioner of state weights and measures and commissioner of the State Liquor Commission. Usually laconic in manner but capable of forthright candor in unmistakeable Yankee tones, he brought political skills to New England's cause at the time when they were most needed. An unabashed foe of the earlier federal efforts, White came to be utterly convinced of the necessity of joint water resources planning on a regional basis, a position regarded by some of his associates as at least ten years ahead of its time.

William S. Wise, chief engineer of the Connecticut State Water Resources Commission, was close to the antithesis of White, yet the two became close friends and effective allies in New England's various water struggles. An engineering graduate of MIT (1923), Wise spent the bulk of his professional career with the commission, rising to become, first, chief engineer and then director of an agency that grew to embrace virtually all of Connecticut's water resources functions. Effective in a quiet but firm way, Wise enjoyed the respect and affection of his professional colleagues for his invariably thoughtful and courtly manner.

Philip Shutler, director of the Vermont Water Conservation Board, began a long and illustrious career in the early 1930s as a member of the Vermont State Planning Board, rising to the post of director of the Water Conservation Board when it was created by legislative act in 1947. His was the central professional voice in the early debates over flood control, and his two decades of determined effort in this regard made him the logical man to serve as director of the Connecticut River Valley Flood Control Commission when the interstate compact was finally approved by Congress in 1952. Small in stature but large in purpose, Shutler never retreated from his stance of states' rights advocate. This led him to harbor some doubts in later years as relations grew increasingly cordial between state and federal officials.

By the time of NRC, state leadership was firmly in the hands of White of New Hampshire and Wise of Connecticut. However, the director and chief engineer of the Massachusetts Water Resources Commission, *Clarence I. Sterling, Jr.*, an internationally respected sanitary engineer, was also a force to be reckoned with. Upon Sterling's untimely death, the chairman of the commission, Massachusetts commissioner of natural resources *Charles H. W. Foster*, became an active member of the NRC group. Interest was growing in northern New England as well. *Richard W. Macomber*, special assistant to Vermont Governor Hoff, and *Rheinhold Thieme* of the Vermont Water Resources Board, alternated as the official Vermont representatives during NRC. At Foster's suggestion, Maine's forest commissioner *Austin H. Wilkins* was drawn into the discussions. As an architect of the interstate compact on forest fire protection, Wilkins had established regional credentials. In fact, in the delicate period of transition from NRC to NERBC, when the states' position was most critical, it was Macomber, Foster, and Wilkins who supplied most of the leadership.

On the federal side, the Corps of Engineers played a leading role throughout the entire study period. Much has been said already about *Col. F. F. Frech* and *Col. Benjamin Talley*, the North Atlantic Division engineers during the NENYIAC period, but credit is also due New England Division engineers *Brig. Gen. Robert J. Fleming* and *Col. Remi O. Renier* for helping keep the regional program alive during the NRC era. Agriculture's *Alvin C. Watson*, a veteran Soil Conservation Service employee, also supplied crucial leadership during NRC, as did *Sylvan C. Martin*, the northeast regional engineer of the Public Health Service. From a prominent position during NENYIAC, the Federal Power Commission's role declined as the issue of public versus private power receded in importance. Nevertheless, the FPC's *John H. Spellman*, a former New Hampshire Water Resources Board engineer, found himself intimately involved in regional affairs for more than two decades, closing out his public service career as director of NERCOM's New England Energy Policy Project.

A special word is owed Interior's *Mark Abelson*, whose span of service extended through the early days of NERBC. Interior's New England relations were first assigned to its Division of Power but later transferred to its Program Staff, a Washington policy and program unit reporting directly to the Secretary. Interior's official representatives on NENYIAC and NRC were the chairmen of the Boston-based Northeast Field Committee, a unit of the Program Staff whose function was to coordinate the work of all department bureaus within the region. As Northeast regional coordinator, Abelson survived the early wars over public power, plunged into combat with the Corps of Engineers over the impact on fish and wildlife of the proposed NENYIAC reservoirs, advanced to the extent he could the states' interest in a federal-interstate compact, and facilitated the NRC and NERBC operations until separate office and staff facilities were available. Small in stature, he was long in interagency experience, and Abelson's steady participation over the years is a central part of the New England river basins story.

The first chairman of NERBC was *R. Frank Gregg*, a native Coloradoan and veteran conservation professional with extensive service in Washington in both the public and private sectors. A suave and savvy individual, Gregg brought to the job both national prestige and extensive Washington connections in and out of the bureaucracy. For example, his prior service as executive director of the Citizens Committee for the Outdoor Recreation Resources Review Commission Report had given him close ties to Congress. His friendship with Henry Caulfield, the first director of WRC, guaranteed NERBC an attentive ear in the federal interagency structure.

Gregg was an innovator with sharp, clear objectives for his fledgling regional commission. He had the knack of listening to complex discussions and then summarizing accurately the most productive courses of action. He got along well with people and was particularly admired by his staff. In moments of stress, he would take off his glasses, close his eyes, and appear to take a direct message from God.

Gregg's occasional detractors found him dominant and doctrinaire at times. He could both threaten and annoy others. Despite discussions to the contrary, the minutes of commission meetings would invariably reflect a course of action perceived by the chairman to be in the best interest of his agency. Nevertheless, Gregg was a strong, effective chairman—the unquestioned dean of all the river basin commission chairmen—and, in the opinion of one veteran observer, "one of the five or six best public servants I have ever known."[92]

When Gregg left in 1978 to head Interior's Bureau of Land Management, his successor was *Dr. John R. Ehrenfeld*, long-time consultant with Arthur D. Little, Inc. and the founder of the environmental firm of Walden Research, Inc. in Cambridge, Massachusetts. Ehrenfeld had mentioned to Massachusetts Lt. Gov. Thomas O'Neil his interest in a public service assignment. When the jobs of Massachusetts secretary of environmental affairs and regional director of the Environmental Protection Agency went to other individuals, O'Neil and Gov. Michael Dukakis recommended Ehrenfeld for the vacant NERBC chairmanship, an assignment he later came to regard as a highlight of his professional career.

Ehrenfeld's task was not an easy one, inheriting as he did the mystique of Frank Gregg. Nevertheless, his skills were those the commission needed badly, the structured and product orientation of a consulting engineer. It was not long before Ehrenfeld had streamlined staff responsibilities, consolidated the commission's unwieldy committee structure, and brought its members more into the mainstream of the agency's planning activities. He found, in his own words, "a lot of dither"[93] in NERBC's program, objectives, and goals. It was time to establish a set of institutional objectives and reduce the episodic nature of its program activity.

In accomplishing change, Ehrenfeld was commanding and forceful, often impatient with the process of subtle interaction that had marked the Gregg years. His authoritarian presence, as one observer put it, was better fitted to the bridge of a battleship. Yet Ehrenfeld's technical credentials generated immediate respect, and he was a person invariably comfortable to be with. The state members, in particular, felt the new chairman more responsive to their interests. He had genuine state concerns but still insisted that the commission's state projects operate within regional themes. Had there been more time, Ehrenfeld's analytical and leadership qualities might even have solved the enigma of all the river basin commissions—an operable definition of the comprehensive, coordinated, joint plan (CCJP) required by law.

In *Robert D. Brown*, the staff director of the commission for all but the first four years of its life, both chairmen had the benefit of a talented and dedicated planning professional. Brown's acquaintance with the commission began during his service as director of Connecticut's Capitol Region (Hartford) planning agency. Among his concerns was the possible diversion of Connecticut River water to Boston via the Northfield Mountain (Massachusetts) pumped storage project.

When Brown came on board in 1971, he found himself the titular head of a sprawling planning enterprise. In fact, he came to do what the chairmen chose not to do themselves, the burden of working for strong, program-oriented chairmen. There were regular staff meetings, but the quarterly commission meetings provided the best vehicle for internal coordination. As NERBC's studies flourished, so did Brown's reputation as a capable and tenacious staff director on the cutting edge of most of the significant decisions. He provided another quality of incalculable benefit to the commission over the years, an institutional memory.

When the commission's demise was imminent, it was Brown who

stepped forward to try to save the day, initiating staff studies of possible successor organizations and rallying the support of the citizens Steering Committee. Brown's "end-run"[94] generated a measure of ill will from NERBC's state members. Nevertheless, when the new New England–New York Water Council was formed, these same members turned to Brown again for consulting advice on organization and program.

Of the numerous federal members and alternates over the years, two individuals performed important service as interregnum chairmen. The first was *Col. John P. Chandler,* New England Division Engineer of the Corps of Engineers, who served for four months until John Ehrenfeld succeeded Frank Gregg as chairman. The second was *Comdr. Stephen L. Richmond* of the Coast Guard who had the unenviable job of closing out the commission when it was terminated in 1981. Unlike the other river basin commissions, NERBC chose not to utilize the vice-chairman as interim chairman, because it interpreted the chairman's job as being that of the leader of the federal agencies.

Colonel Chandler recalled a measure of surprise when Frank Gregg asked him to serve as alternate chairman, for he and Gregg had drawn swords upon occasion over regional issues. Despite these differences, Chandler's interest in New England was well known, and he was flattered to have been asked to perform even a caretaker function. There was substantial institutional significance as well, for in many regions of the country the Corps of Engineers was not held in high regard by state and federal members. Gregg also knew that the incoming Carter administration had consulted with Chandler about NERBC. Chandler had advised presidential advisor Jack Watson that the commission was clearly the best of the regional agencies in New England.

From service as alternate to the commanding admiral of the First District of the Coast Guard, the official NERBC member for the Department of Transportation, alternate chairman Stephen L. Richmond went from relative obscurity to sudden prominence. It was his sad duty to process the personnel termination notices and close the doors of an organization he had come to regard highly. Richmond's selection as acting chairman also caught him by surprise. The interests he shared with chairman Ehrenfeld in regional issues and a better definition of the CCJP appear to have marked him as the logical successor on the federal side. Once appointed, he gave the assignment characteristic vigor and attention.

For example, in an effort to "return power to the people,"[95] as he put it, Richmond encouraged the staff to activate the citizens Steering Committee and explore options for program continuation. His letter to the governors in April 1981 was warmly supportive of the commission. He had three primary objectives in assuming the chairmanship: to oversee the commission's orderly termination, to ensure that it completed its business respectably, and to provide some kind of carry-over to a successor program. All three objectives had been achieved when NERBC closed its doors on September 30, 1981.

On the states' side, fourteen individuals served as vice-chairman or alternate vice-chairman during the fourteen-year life of the commission. Each was elected annually, and some served for more than one term. After 1975, the practice was to elect the state member whose governor was serving as co-chairman of the New England Regional Commission in order to encourage a measure of collaboration between the two regional agencies and attract the direct interest of the New England governors.

The job of vice-chairman was not arduous, consisting primarily of presiding over the caucus of state members held just before each quarterly meeting. In addition, the vice-chairman had certain approval and advisory functions and, ultimately, in case of disagreement between state and federal members, the statutory responsibility for setting forth the official position of the states. Unlike the chairman, the vice-chairman had his regular duties to attend to back home. The staff assigned specifically to him by the commission were limited. In later years, a specific budgetary item did allow for expense reimbursement for any state member officially representing NERBC such as testifying at congressional hearings.

Despite these limitations, the assignment of vice-chairman was taken seriously by the state members and, in a number of instances, real leadership was provided. *Austin H. Wilkins,* for example, Maine's forest commissioner at the time of NERBC's formation, worked closely with chairman Gregg on organizational matters. Vermont's *Lemuel Peet* contributed substantial leadership in guiding the commission toward its first regional priorities reports. Massachusetts's commissioner of natural resources *Arthur W. Brownell* served the longest period as vice-chairman – three successive terms – finding the NERBC meetings, on the whole, interesting and useful.

To Vermont's *Arthur Ristau,* the state planning director, NERCOM executive director, and NERBC state member, fell the delicate task of presiding over the first joint venture with the New England Regional Commission. Under Ristau, the experiment was largely successful; successive NERCOM alternates saw less value in NERBC's program. Yet closer contact with the governors could be useful. Under Massachusetts's *Evelyn Murphy's* vice-chairmanship, for example, the commission developed a New England policy position on national water policy that was reinforced through the National Governors Association.

As always, there were other influential state representatives whose names do not appear on the official roster of officers. The state planning director from Rhode Island, *Daniel Varin*, gave steady counsel and support throughout the commission's existence. His analysis of alternative institutional arrangements for New England was particularly timely. Connecticut's *John Curry* and *Hugo Thomas* never hesitated to let NERBC know where they stood on issues. And to Vermont's *Bernard Johnson*, New Hampshire's *Peter Piattoni*, and Massachusetts's *Elizabeth Kline* fell the difficult task of engineering a transition from the old NERBC to the new New England–New York Water Council.

In Summary

From NENYIAC to NERBC, the bottom line on leadership has to be outstanding regardless of agency affiliation. The institutional fabric was so fragile throughout much of the period that, without individual commitment, nothing much would have been accomplished. What was particularly noteworthy was the effect of New England's magic on even experienced agency bureaucrats. Special relationships developed between state and federal officials that helped sustain their institutions even in moments of stress, and a genuine sense of regional commitment emerged that constitutes an enduring heritage. Small wonder, then, that of all its offspring, WRC staff regarded New England as the top river basins commission in the country – an agency with a militant sense of independence but significant creativity in its pursuit of public policy.

THE PRIVATE SECTOR

A curious feature of the NENYIAC, NRC, and NERBC periods is the limited extent of participation by the private sector. Yet, for much of the time, New England was riding the crest of a national ground swell of interest in natural resources and environmental matters.

In the early days of NENYIAC, and particularly during the initial round of hearings, spokesmen for private organizations appeared publicly to register complaints about hydroelectric power and voice fears about prospective valley authorities. The statements were usually charged with emotion, and based primarily on hearsay. For example, the December 1954 newsletter of Wildlife Conservation Inc., a small private conservation organization headquartered in Boston, stirred NENYIAC's particular indignation by referring to the newly released recommendations as New England's "Christmas cravat, bright and full of promise, but a little tight around the neck!"[96]

During the course of the NENYIAC studies, Corps of Engineers officials seem to have received only two formal invitations to appear before major conservation groups: one from the New Hampshire Natural Resources Council in 1954, whose promotional circular asked the
question: "What Has Happened To The Great New England Resources Survey That Was Started Several Years Ago?"; the other from
the Rhode Island Wildlife Federation in 1955, which used a series of
speakers from NENYIAC as the drawing card for its annual meeting
program.

During the second round of hearings, most of the private conservation spokesmen centered their opposition on projects believed to en-
danger either the New York Forest Preserve or the fish and wildlife
resources of individual river basins. The participants, however,
showed little real knowledge of, or interest in, NENYIAC's overall objectives. In fact, had it not been for the missionary work of two national conservation organizations, the National Wildlife Federation
and the Wildlife Management Institute, each with field representatives located in New England, and each primarily interested in wildlife, few spokesmen would have appeared from the private conservation sector at all.

The unhappiness over fish and wildlife resource aspects at NENYIAC-
identified reservoir projects could be traced directly back to Interior's
losing effort at promoting alternate sites, a fact that was quietly re-
layed to the agency's supporters on the outside. Strangely, however,
Interior's most influential constituency, the various state fish and
game agencies, played no really prominent role in the dissent. Such
leading regional figures as New Hampshire's Ralph Carpenter and
Maine's Roland Cobb actually praised NENYIAC publicly for the stimulus it had given long-overdue fish and wildlife evaluations.

During NRC, little improvement was noted in regional public relations despite a recognition on the part of earlier NENYIAC participants
that its public image could stand some improvement. The state meetings were, in part, a sincere effort to remedy the situation, but even
when NRC sessions were thrown open to the public and special presentations were arranged on particularly timely topics, the attendance outside of NRC's regular agency clientele and loyal band of
camp followers was miniscule at best.

Once the earlier and more controversial aspects of the regional
study had faded from view, New England's news media carried only
the briefest of formal notices concerning either NENYIAC or NRC.
Outside of times of floods and drought, their water resources deliberations were just not news. Why this should be the case is worth pursuing for a moment.

What was really lacking in New England during this period was a good way of bridging the gap between the public and private sectors. One promising possibility, the private watershed movement, was still in its infancy and severely handicapped by the multitude of political boundaries it was forced to recognize and the limitations of its all-embracing philosophy. Thus, NENYIAC and NRC were denied the feedback they so badly needed from the private sector, and the private organizations lacked the accurate information essential to real public understanding.

By the time of NERBC, however, much had changed for the better. A growing environmental awareness had given water resources a surer seat at the natural resources table. It had now become fashionable to take a broad view of the field, and the agencies were responding to this quickening public interest by building their own bridges to these new and promising constituencies.

Not surprisingly, NERBC left little time before undertaking a substantial public information program. Its studies and annual reports were skillful productions. A monthly newsletter sprinkled the region with facts about water—and about the commission. As a former wildlife information officer, chairman Gregg cared about good writing and good reporting. In the early days of NERBC, the chairman and staff roamed the region addressing audiences on the topic of the new commission. Road shows were arranged for each of the state capitals. An imaginative feature of later years was the topical workshop sponsored by NERBC in conjunction with most of its quarterly meetings. The commission was also quick to hold conferences on timely issues, often in cooperation with other organizations. Examples were the regional economic conference cosponsored with the New England Regional Commission in 1974, and the postaudit session on coastal flooding held after the severe winter storm of February 1978. Other successful ventures were the film *Offshore/Onshore,* produced by station WGBH (Boston) in 1977 for the public broadcasting network, and, in fact, most of the commission's products from its special studies of the Georges Bank offshore oil issue.

But despite these impressive projects, and the favorable disposition of the public to such topics during the environmental decade of the 1970s, NERBC simply did not manage to build a constituency for itself. In the light of the elaborate provisions for public participation in most of its comprehensive planning investigations, the absence was more than accidental. In trying to serve all the needs of the region, NERBC felt that it could not afford to become captive of any single group.

At the end, chairman Ehrenfeld recognized the commission's vulnerability and moved to constitute a citizens Steering Committee on

Public Involvement. The effort was merely tolerated by the members because of painful experiences of their own with public participation. Many were still smarting from the well-intentioned but unmanageable citizen committees in the Connecticut Valley. But the wisdom of this move paid off in the end. The Steering Committee, nebulous though it was, was the lone outside voice favoring a successor regional water resources agency.

As for the environmental institution constituency, much of what the commission set out to do did not mesh well with citizen organization priorities. The leading state organizations were concerned with natural area preservation, forestry, recreation, and environmental protection—not comprehensive planning. It was not until the later studies of water quality and hydro power that a natural community of interest developed. And many of the state-level organizations had become frankly disenchanted with regionalism, seeking other dragons to slay. The only natural conservation constituency, the watershed association, remained scattered and preoccupied with local concerns. By 1977, the regional organization most likely to "shadow" NERBC, the New England Natural Resources Center, had closed its staffed offices in Boston and reverted to a trustee structure only.

One other potential constituency group was business and industry, long a potent political force in New England affairs. Prominent among such interests was the New England Council, the region's self-styled "town meeting of New England business that never adjourns." The council was the outgrowth of a New England conference of business and industry leaders, called by the New England governors for November 1925, to help offset the steady move of the textile industry southward. In 1937 the council returned the favor by advancing the suggestion that the governors themselves organize for the purpose of regional cooperation. For many years, the council provided the secretariat for NEGC. Organized into six eighteen-member state councils, the council's program developed in four principal areas: interstate relations, economic action planning and research, vacation travel–natural resources, and Washington relations.

The council appears to have been an active, if behind-the-scenes, participant during the early days of NENYIAC. It was the council's Power Survey Committee, for example, that first challenged the figures on hydroelectric potential advanced by the FPC. It was also the council that took an active interest and finally published the Committee of New England report on the general state of the economy in New England.

Once the initial uncertainties over NENYIAC's course were settled, there is reason to believe that the council succumbed to the prevailing public opinion that NENYIAC would be just another governmental study. General liaison was maintained by executive vice-president Dudley Harmon, but the council did not participate directly in any of the NENYIAC work groups.

Harmon's growing personal interest in NENYIAC, as revealed by the series of articles he wrote for the *Providence Journal* in 1955, plus the advent of the 1955 floods, seem to have rekindled the council's interest in regional river basin affairs, as evidenced by its endorsement and active pursuit of a New England flood relief program.

Consequently, when NRC chairman Walter White approached Harmon's successor, Gardner A. Caverly, for assistance late in 1957, Francis E. Robinson of the council's interstate relations staff was not only assigned to help but given instructions to maintain continuing liaison with NRC.

Robinson, an agricultural information specialist and former assistant to the president of the University of New Hampshire, grew genuinely interested in the NRC cause. His willingness to go to bat for the organization within the council's administrative staff and committee policy structure seems to have been among the important factors leading to the council's endorsement of the interstate compact and its willingness to assume the secretarial burden when state representatives succeeded to the chairmanship.

NRC, in turn, was more than helpful to the council's natural resources program, which had never been strong. Debate within the council over endorsement of the compact led to a review, and later consolidation, of committees under the single head—natural resources—with a staff member of the council assigned for full-time guidance.

The compact issue also gave the council another good reason to launch its much-discussed Washington office early in 1961, the unsuccessful efforts of 1960 revealing that a full-time representative in the nation's capital was bound to be far more effective in advancing New England's various causes.

Having patched up its differences with NEGC in 1958, the council could also lend an effective hand with New England's political leadership. Melvin Peach of the council, acting again as secretary of the Governors Conference, was in a key position to keep the compact issue in focus through his day-to-day contacts with the governors. This gave NRC a badly needed hand in an area in which it had always been weak.

Robinson's replacements, Peter C. Janetos and Edwin Webber, proved no less effective than their predecessor. Webber, in particular, a former University of Rhode Island government professor, utilized his knowledge of interstate relations and his excellent contacts in southern New England to advance the NRC cause materially. It was Webber's skilled assistance, for example, that produced the first drafts of New England's projected river basins organization under the provisions of the Water Resources Planning Act. Janetos and Webber also served as continuing bridges between the New England state planning directors and NRC, an effort that did much to convince the governors that natural resources planning was being properly coordinated in their region.

Outside of directly water-related concerns, such as those of the utilities, New England industry played no discernible role in NERBC. It viewed the commission as environmentally oriented, and its normal regional relationship was to the New England Regional Commission, an agency with economic concerns as its primary mission. But the fact of the matter was that the industry's leading regional organization, the New England Council, was in some disarray during much of the NERBC period. The council's logical point of interface, its committee on natural resources and tourism, was largely inactive and later abolished outright.

In retrospect, the council's involvement in river basin affairs over the years has been sporadic at best, and peripheral at worst. Although substantial and important contributions have been made by the council, they were advanced largely by individuals—staff associates and key committee members—not the council as a whole. Natural resources issues, to this day, have not enjoyed a priority position in the council's program.

In Summary
The inescapable truth is that New England's river basin affairs over the years have been largely intergovernmental ventures. The private sector's role has been more one of reaction than of action. This has left the agency and political actors free to play their own game without the counterweight of a focused public opinion. The positive force of accountability has been lacking. What is ironic, however, is that when the private sector has been enlisted in an important cause, such as the struggle for the northeastern resources compact, an important ally was added. In a region as resource-dependent as New England, it is curious that a corresponding regional constituency has not developed to help guide the governmental efforts.

CONCLUSIONS
What then are the lessons to be learned from this thirty-year course of river basin history? There are several that can serve as useful guidelines for subsequent institutional approaches in New England.

The New Englander's roots run deep, as the record has indicated. The principles of representation and participation are central to his interest and support, drawn from a centuries-old tradition of self-reliance and self-determination. Regardless of its limitations, the town meeting approach still flavors New England attitudes, and any proposal must run the full gamut of public scrutiny if it is to succeed and endure. For so human an environment, the *whos* of an issue are far more likely than the *whats* to become the central question. Thus the credibility of the proponents, not the validity of the proposals, may prove the final determining factor in any given issue.

Though pragmatic by nature, the native New Englander is cautious in accepting a new approach. He demands a suitable gestation period, accepting proposals from within with greater frequency than from without. Once convinced of its merits, however, his dedication to accomplishment is utter and complete.

There is also a certain frugality and thrift in the New England judgment. Issues will often be weighed not by what is to be gained, but more by what may be lost. Call it caution, conservatism, or even provincialism—this attitude tends to militate against the sudden venture, the fanciful whim, the careless action.

To the outsider, New England often appears to operate in an atmosphere of seeming anarchy where the lowest common denominator offers the best chance for success. A careful balancing of interests *is* often the key to accomplishment but only as a fine timepiece with its many moving parts can become a priceless instrument. From its many voices and its pluralistic approaches, New England derives its traditional strength.

New England's next venture, the New England–New York Water Council, must be sensitive to these historical realities, yet modern in its recognition of new opportunities and needs. As an example, virtually all of the New England states have now moved toward consolidated natural resource and environmental agencies, often at the cabinet level of government, an event that should materially reduce the number of actors to be dealt with on a regional basis. Emergence of this new governmental form, at least in theory, provides readier access to the chief decision makers, the governors. Strengthened state agencies and functions have also reduced the once dominant role of the federal agencies. For the first time in recent history, the real key to bioregional accomplishment may rest with the states' ability to work together on matters of mutual interest. Any bioregional institution formed for such purposes must itself be different: respectful of its constituents, modest in its aspirations, flexible in its approach to problems, and always mindful of the need for change. Above all, it must be blessed with a sixth sense of what is timely and doable, and exercise at all times the high political art of the possible.

But what does the New England experience say about bioregionalism in general? There are larger conclusions to be drawn that seem worth addressing in these closing paragraphs.

First, a cardinal principle for most bioregionalists has been the matter of ecological integrity. It is considered important, for example, that the entire watershed of a river system be included within the jurisdiction of the managing agency. In New England's case, WRC insisted that New York be included within the initial NERBC region. The agency itself, in later years, expended considerable time and political capital on the needs of a state that never really felt it belonged to the region. Far more important than technical ecological integrity would seem to be a sense of regional belonging on the part of people living within the area. Without that essence of regional consciousness, no bioregional entity has a chance of succeeding.

The second essential ingredient for a viable bioregion is the inclusion of whole political jurisdictions, preferably on a scale of approximate power parity. Thus the six New England states could work together cooperatively despite their differences in size and degrees of affluence because they were similar political entities. Including a portion of a political jurisdiction just for the sake of ecological or structural integrity is a likely prescription for trouble. A participant may be gained, but one lacking a concomitant degree of commitment to the larger regional goals.

Third, the scale of a prospective bioregional entity is worth examining carefully. It must be large enough to dwarf jurisdictional differences, yet small enough to encourage focused, implementing actions. It must have sufficient critical mass to be economically and politically viable, yet also seem manageable to its participants. Questions of scale apply equally to the bioregional institution itself. For example, it must have enough staff and budgetary support to function effectively, yet not be so large as to represent a drain on or a threat to its constituent parts. The New England effort was significantly flawed in each of these respects.

How a bioregional institution comes into being can also be important. Ironically, most successful bioregional institutions are more often products of fortuitous or compelling circumstance than of careful design. Building from the bottom up, preferably on a previous base of mutual interest and cooperation, is infinitely preferable to a mandate from on high. There should also be clear and compelling reasons for its creation, and those premises should be examined at regular intervals to prevent the inevitable process of drift and decay.

In New England's case, the genuine roots present at the outset became subsumed by the predominantly federal goals of the resultant agency.

What the organization does obviously becomes the primary determinant of its viability. Its program must be well defined, but not rigidly proscribed. It should be built around specific, timely problems or issues, yet contain the capacity to expand or contract to meet needs as they occur. Such generalized functions as information exchange, regional coordination, and regional policy effectuation are preferable to roles as an operating or managing agency, which tend to lock the regional organization into functions that others could perform and to harvest needless amounts of ill will. In general, it should aim to provide its constituents with something over and above what they can supply themselves—technical specialists, brainpower, data, funds, the capacity to mediate, the ability to influence decisions. NERBC, regrettably, had just gone through a cleansing reformation when it was abruptly terminated by the new administration.

What's next for bioregionalism? A rich and exciting future, in my opinion. There is an urgent need for further experimentation, and significant opportunities in places like the Great Lakes and Chesapeake Bay, to name just two. The principle of bioregionalism can be applied literally anywhere a regional problem exists. But there is a need to look critically at other areas with significant regional experience, such as Alaska, the bay area and desert regions of California, Lake Tahoe, and the Colorado River Basin in the west, and, to name obvious candidates in the east, Appalachia, the Adirondacks, the Ozarks, and the Ohio, Delaware, Susquehanna, and Potomac river basins.

The need for serious attention to bioregional approaches and institutions is heightened by the likely reduced federal presence for at least the balance of this century. In that event, it will be squarely up to the leadership of the states to devise new ways to bridge jurisdictional differences so that natural resources and environmental matters can be addressed in the context within which they occur naturally.

Appendix of
Water Resources Documents

FLOOD CONTROL ACT OF 1950
(Chapter 188, Public Law 516)
Approved May 17, 1950
Sec. 205. The Secretary of the Army is hereby authorized and directed to cause preliminary examinations and surveys for flood control and allied purposes, including channel and major drainage improvements, and floods aggravated by or due to wind or tidal effects to be made under the direction of the Chief of Engineers, in drainage areas of the United States and its Territorial possessions, which include the following-named localities, and the Secretary of Agriculture is authorized and directed to cause preliminary examinations and surveys for run-off and water-flow retardation and soil-erosion prevention on such drainage areas, the cost thereof to be paid from appropriations heretofore or hereafter made for such purposes: *Provided,* That after the regular or formal reports made on any examination, survey, project, or work under way or proposed are submitted to Congress, no supplemental or additional report or estimate shall be made unless authorized by law except that the Secretary of the Army may cause a review of any examination or survey to be made and a report thereon submitted to Congress if such review is required by the national defense or by changed physical or economic conditions: *And provided further,* That the government shall not be deemed to have entered upon any project for the improvement of any waterway or harbor mentioned in this title until the project for the proposed work shall have been adopted by law:

Merrimack and Connecticut Rivers and their tributaries, and such other streams in the States of Maine, New Hampshire, Vermont, Massachusetts, Connecticut, and Rhode Island, where power development appears feasible and practicable, to determine the hydroelectric potentialities, in combination with other water and resource development.

Arkansas, White and Red River Basins, Arkansas, Louisiana, Oklahoma, Texas, New Mexico, Colorado, Kansas and Missouri with a view to developing comprehensive, integrated plans of improvement for navigation, flood control, domestic and municipal water supplies, reclamation and irrigation, development and utilization of hydroelectric power, conservation of soil, forest and fish and wildlife resources, and other beneficial development and utilization of water resources including such consideration of recreation uses, salinity and sediment control, and pollution abatement as may be provided for under Federal policies and procedures, all to be coordinated with the Department of the Interior, the Federal Power Commission, other appropriate Federal agencies and with the States, as required by existing law: *Provided,* That Federal projects now constructed and in operation, under construction, authorized for construction, or projects that may be hereafter authorized substantially in accordance with reports currently before or that may hereafter come before the Congress, if in compliance with the first section of an Act entitled "An Act authorizing the construction of certain public works on rivers and harbors for flood control, and other purposes," approved December 22, 1944 (58 Stat. 887), shall not be altered, changed, restricted, delayed, retarded, or otherwise impeded or interfered with by reason of this paragraph.

PRESIDENTIAL EXECUTIVE ORDER — OCTOBER 9, 1950
ESTABLISHMENT OF A NEW YORK–
NEW ENGLAND INTERAGENCY COMMITTEE

The White House
Washington, D.C.
October 9, 1950

My dear Mr. Secretary:
You will recall that in connection with my approval of H. R. 5472, the
Rivers and Harbors and Flood Control Acts of 1950 (P. L. 516, 81st Cong.),
I sent a message to the Congress indicating what I considered to be serious
deficiencies in the legislation. Along with other observations, I pointed out
the failure of the measure to provide for a comprehensive study of
multiple-purpose resource development for the New England–New York
area with appropriate participation by the Federal agencies and the States
concerned. The Congress has been considering legislation to meet this deficiency, but final action has not been taken on a bill establishing a commission to conduct the desired study.

I am sure you will agree with me that experience in natural resources
development emphasizes the fact that plans for the most effective utilization of water resources must take into account all the multiple-purposes
and benefits and also the interrelationships between water and land
resources. Moreover, studies of the potential development of these related
resources should be based on proper geographical or regional areas. It was
for these reasons among others that, on February 9, 1950, in a communication to the Vice President, I endorsed the legislation, recently considered in
hearings before a Subcommittee of the Senate Committee on Public Works,
providing for the establishment of a New England–New York Resources
Survey Commission and authorizing a full-scale investigation of multiple-
purpose resource development with participation by the Federal agencies
and the States concerned.

In order to realize to the greatest extent possible under existing authority
the benefits which would stem from this legislation and to provide essential coordination of the activities of the various Federal agencies in studying
the resources potential of this area, I am requesting that the various Federal
agencies concerned, including your Department, organize a temporary
interagency committee for the purpose of initiating a comprehensive survey
of the resources of this region, and preparing recommendations for the
development, utilization and conservation of these resources. In view of
the general provisions of section 205 of Public Law 516, I am designating
the Department of the Army as the Chairman agency. The survey to be
conducted by this interagency committee should include the six New England States and New York State. Of course, the committee can exclude
from the survey any parts of this area whose resources are not well suited

for consideration in a general survey of this kind. The committee should take into consideration the resources of the areas in this which are of mutual interest to the United States and Canada, such as the Passamaquoddy Bay, with due regard to pertinent international agreements between the two countries. You will recognize of course, the responsibility of the Department of State in these matters. That Department, therefore, should be consulted on issues affecting these areas.

In serving on this committee, each agency should make its contribution under existing laws and in accordance with its responsibilities under such laws. And it is most important that the efforts of the various agencies be integrated from the very beginning of the investigation if the benefits of all coordination possible under existing law and procedures are to be obtained. The final product of the interagency survey should be a single comprehensive report setting forth the coordinated findings of all the participating agencies.

Each agency and the committee as a whole should coordinate its plan and activities with those of interested State and local agencies. The State and local agencies have a direct and vital interest in the conduct of this investigation and the report that will result. Many of the existing activities of State and local agencies in the resources field should fit into the projected investigation. I am sure that State and local agencies will be anxious to cooperate. In order that they be afforded every opportunity and encouragement to participate in the work of the committee, I am asking the Governors of the seven States concerned to designate official representatives to act as liaison between the committee and the various State agencies concerned with resource development. In addition, I want the committee to invite the ideas and help of local governments and private groups and individuals to the extent possible. It is essential that the Federal agencies draw upon the experience and ideas of the people of the region to the fullest extent and that the final report carry the concurrence or comments of each affected State.

The comprehensive study of land and water resources of this area should include, among other matters, coverage of electric power generation and transmission, forest management, fish and wildlife conservation, flood control, mineral development, municipal and industrial water supply, navigation, pollution control, recreation, and soil conservation. The necessary first step in such a study is an inventory of the land, water, and all of the related natural resources available for utilization, together with a survey of the projected regional and national requirements which might be met through more effective utilization of the natural resources of the region. When these basic facts on resources and needs have been collected and analyzed, the committee should then proceed to determine what development and conservation projects are feasible and desirable, and to prepare recommendations for specific action to carry them out.

It is my desire that this survey be undertaken as soon as practicable and that the joint report be submitted for my consideration not later than July 1, 1952. In taking part of this investigation, each agency should utilize, to the fullest extent possible, funds available for the fiscal year 1951. It will also be necessary for each agency to furnish immediately to the Bureau of the Budget its budget estimates for fiscal year 1952 for participation in the comprehensive survey. As its first order of business the committee, through joint planning of all the member agencies, should prepare a detailed program spelling out the method by which the comprehensive survey will be undertaken, together with a consolidated statement giving more precise estimates of the fiscal requirements of each agency. These program plans should be submitted to the Bureau of the Budget early next spring so as to permit such budgetary modifications as may appear appropriate at that time.

You will recall that in my letter of July 21, I requested several agencies to conduct a detailed review of their programs for the purpose of modifying them wherever practicable to lessen the demand upon services and commodities which, in view of the present international situation, are needed for national defense. The strengthening of our defense program may delay full implementation of the findings of the proposed survey. Nevertheless, I feel that this initial investigation should go forward immediately since, in providing the blueprints for the most effective development of the resources of this important area of our country, the survey should encompass projects which can make significant contributions to our national defense effort.

There are, of course, certain projects in the general area to be covered by this survey which are plainly good investments for the future of the region and the nation and should be started as soon as possible. Among these are the further development of the Niagara Falls power potential and the construction of the St. Lawrence seaway and power project. The importance of these projects to the national defense make it more necessary than ever that their construction be undertaken immediately. The interrelationship of these projects and their relation to other resource development work in the Northeast should be considered by the inter-agency committee, but this survey should not delay the building of these projects.

Finally, as I have pointed out from time to time, the economic growth and stability of an area depends largely upon how its natural resources are developed. Last spring the Council of Economic Advisers appointed a Committee of Experts on the New England Economy to prepare an analysis of New England economic opportunities and problems, which is now nearing completion. It is highly desirable that the findings of this Committee be taken into account in developing the report of the interagency committee.

Identical letters are being sent to the Departments of the Interior, the Army, Agriculture and Commerce, and to the Federal Security Agency and the Federal Power Commission, the agencies which will participate as members of the committee. I am also forwarding a copy of this letter to the Department of State.

Sincerely yours,

SIGNED: Harry S. Truman

INTER-AGENCY COMMITTEE ON
WATER RESOURCES – JUNE 29, 1956

Charter for a Northeastern Resources Committee

1. *Purpose*

It is the purpose of this agreement to provide in the Northeastern region improved facilities and procedures for the coordination of the policies, programs and activities of the States and Federal agencies in the field of water and related land resources investigation, planning, construction, operation, and maintenance; to provide means by which conflicts may be resolved; and to provide procedures for coordination of their interests with those of other Federal, local governmental, and private agencies in the water and related land resources field.

2. *Establishment*

(a) For this purpose there is established a Northeastern Resources Committee of State and Federal representatives operating on a basis of co-equality. The Committee shall be composed of representatives of any of the following States and Federal agencies which indicate a desire to participate: The States of Maine, New Hampshire, Vermont, Rhode Island, New York and Connecticut, the Commonwealth of Massachusetts, and the Federal Departments of the Interior; Commerce; Labor; Agriculture; Health, Education and Welfare; and the Army; and the Federal Power Commission.

(b) The Governor of each of the States desiring to participate shall designate the member of the Committee for his State.

(c) The Federal members on the Committee shall be designated by the head of the Federal agency they are to represent and shall preferably be resident in the area.

(d) Committee members may designate other officials to serve as alternates.

(e) Federal agencies will participate in the work of the Committee in accordance with their respective responsibilities and interests and with the intent of the "Inter-Agency Agreement on Coordination of Water and Related Land Resources Activities" as approved by the President on May 26, 1954.

(f) When appropriate, other Federal, state, public and private agencies will be asked to participate in Committee meetings and to appoint representatives to specific subcommittees, in order that the work of the Committee may be coordinated with the related work of all agencies.

(g) A Chairman shall be elected annually from and by the State and Federal members; the Chairman shall not succeed himself.

(h) The State or agency which provides the Chairman shall also provide the Secretary for the Committee and the necessary administrative support incident to his tenure.

(i) The Committee shall have such additional staff assistants as the members may, upon request, assign to it.

3. *Method of operation*
   (a) Meetings will be held as often as required, at times and places appropriate to the agenda and normally *at intervals of not more than two months*. Meetings normally will be open to the Public and the Press. Special Executive sessions of the Committee may be held at the call of the Chairman.
   (b) The Committee shall serve as a means for coordinating activities and achieving accord or agreement, at the regional level, among its member States and agencies on issues or problems which may arise. Staff work necessary to coordinate activities and present the essence of any issues or problems to the Committee shall be carried on by the Committee staff or by subcommittee as appropriate and as may be appointed by the Chairman and approved by the Committee.
   (c) Minutes of meetings will be prepared to record the actions and recommendations of the Committee. The minutes will be primarily for use of the participating agencies, but a wider distribution may be made when considered desirable by the Committee.
   (d) The Committee may establish further procedures governing its operations as required.

4. *Responsibilities*
   (a) It will be the responsibility of the Committee to establish means and procedures to promote coordination of the water and related land resource activities of the States and of the Federal agencies; to promote resolution of problems at the regional level; to suggest to the States or to the Inter-Agency Committee on Water Resources changes in law or policy which would promote coordination, or resolution of problems; and in its discretion to communication with the Inter-Agency Committee on Water Resources on any matters of mutual interest.
   (b) The efforts of the Committee on coordination of work and resolution of conflicts will be directed towards all State and Federal activities involved in their respective water and related land resources development responsibilities and shall include coordination of the following:
      (1) Collection and interpretation of basic data.
      (2) Investigation and planning water and related land resources projects.
      (3) Programming (including scheduling) of water and related land resources construction and development.

5. *Geographical jurisdiction*
   The geographical area to be encompassed within the sphere of Committee influence will include the entire states of Connecticut, Rhode Island, Massachusetts, Vermont, New Hampshire, Maine and New York, including Passamaquoddy Bay, Long Island Sound, and the Atlantic Ocean contiguous to the Northeastern Region.

Issued by the Inter-Agency Committee on Water Resources

Department of the Interior
Department of Agriculture
Department of Commerce
Department of Health, Education and Welfare
Department of Army
Federal Power Commission

June 29, 1956   Washington, D.C.

NEW ENGLAND GOVERNORS'
RESOLUTION–MAY 17, 1958

Memorandum of Understanding
Relative to Northeastern Resources Committee

In accordance with their understanding that an informal plan, lacking the attributes and solemnity of a Compact, could be effectuated, those states and agencies participating in such plan could mutually work together for the purpose of improving the resources of the Northeastern region of the United States, it is the intention of the undersigned Governors of the New England States, by subscribing to the plan hereinafter set forth, to indicate the desire and willingness of their respective States to approve, in the following terms, informal plans and procedures relating to the improvement of the resources of the said Northeastern region:

WHEREAS, on the 26th day of May 1954, the President of the United States by letter to the Secretary of the Interior approved the Inter-Agency Agreement on Coordination of Water and Related Land Resources Activities submitted by the Department of the Interior to the Director of the Bureau of the Budget. The purpose of said agreement was to provide improved facilities and procedures for the coordination of the policies, programs and activities of the Departments of the Interior, Commerce, Labor, Agriculture, Health, Education and Welfare, and the Army, and the Federal Power Commission in the field of water and related land resources investigation, planning, construction, operation and maintenance, to provide means by which conflicts may be resolved and to provide procedures for coordination of their interests with those of other Federal agencies in the water and related land resources field. Under this agreement there was created the Inter-Agency Committee on Water Resources; and

WHEREAS, on June 29, 1956, the said Inter-Agency Committee on Water Resources as follows: [see pp. 182–184 for copy of charter].

RESOLVED: That the States of Maine, New Hampshire, Vermont, Rhode Island, Connecticut, and the Commonwealth of Massachusetts, and the United States of America represented by the Department of the Army; Interior, Agriculture; Commerce; Labor; Federal Power Commission; and Health, Education and Welfare; acting through their authorized representatives; being cognizant of the mutual benefits to be derived from the objectives of the above terms, do state it to be the intent of the undersigned, without committing their respective States to any undertaking relative to the appropriation of funds for the benefit of the Committee, and recognizing only the propriety of expenses incidentally involved in connection with attendance at meetings of the Committee of persons regularly employed by the participating States or agencies, to adopt the said terms as being the informal plans and procedure relating to the coordination of the policies, programs and activities of the States, and Federal agencies in the fields of water and related land resources; provided, also, it is the understanding of the signatories that the product of the Committee's work will be in the form of recommendations, which recommendations, insofar as they relate to matters which affect a participating State, may be accepted or rejected by the affected state in the sole discretion of such State.

SIGNED: Joseph B. Johnson
Governor of Vermont,
Chairman, New England
Governors' Conference

Abraham A. Ribicoff
Governor of Connecticut

Edmund S. Muskie
Governor of Maine

Foster Furcolo
Governor of Massachusetts

Lane Dwinell
Governor of New Hampshire

Dennis J. Roberts
Governor of Rhode Island

May 17, 1958.

NORTHEASTERN WATER AND
RELATED LAND RESOURCES COMPACT

Article I: *Findings*

The northeastern part of the United States is by virtue of geographic location and other characteristics a great natural resource area which, with more intense use of natural resources, increasingly requires coordinated planning as a basic ingredient of effective resource management and orderly growth of the region. The work of the New England–New York Interagency Committee demonstrated that a continuation and furtherance of activities such as those undertaken by it would be of great value. To this end, it is the intent of this compact to establish and provide for the operation of a joint agency for the northeast.

Article II: *Purpose*

It is the purpose of this compact to provide, in the northeastern region, improved facilities and procedures for the coordination of the policies, programs and activities of the United States, the several states, and private persons or entities, in the field of water and related land resources, and to study, investigate, and plan the development and use of the same and conservation of such water and related land resources; to provide means by which conflicts may be resolved; and to provide procedures for coordination of the interests of all public and private agencies, persons and entities in the field of water and related land resources; and to provide an organization for cooperation in such coordination on both the federal and state levels of government.

Article III: *Creation of Commission*

There is hereby created the Northeastern Resources Commission, hereinafter called the commission.

Article IV: *Membership*

The commission shall consist of one member from each party state to be appointed and to serve, in accordance with and subject to the laws of the state which he represents, and seven members representing departments or agencies of the United States having principal responsibilities for water and related land resources development to be appointed and to serve in such manner as may be provided by the laws of the United States.

Article V: *Functions*

It shall be the responsibility of the commission to recommend to the states and the United States, or any intergovernmental agency, changes in law or policy which would promote coordination, or resolution of problems, in the field of water and related land resources. The efforts of the commission in coordination of work and resolution of conflicts may be directed towards all state and federal activities involved in water and related land resources development responsibilities and shall include coordination of the following:

1. Collection and interpretation of basic data;
2. Investigation and planning of water and related land resources projects;
3. Programming (including scheduling) of water and related land resources construction and development;
4. Encouraging of the referral of plans or proposals for resources projects to the commission.

The commission shall use qualified public and private agencies to make investigations and conduct research in the field of water and related land resources, but if it is unable to secure the undertaking of such investigations or original research by a qualified public or private agency, it shall have the power to make its own investigations and conduct its own research. The commission may make contracts with any public or private agencies or private persons or entities for the undertaking of such investigations, or original research within its purview.

Article VI: *Voting*

No action of the commission respecting the internal management thereof shall be binding unless taken at a meeting at which a majority of the members are present and vote in favor thereof; provided that any action not binding for such a reason may be ratified within thirty days by the concurrence in writing of a majority of the commission membership. No action of the commission respecting a matter other than its internal management shall be binding unless taken at a meeting at which a majority of the state members and a majority of the members representing the United States are present and a majority of said state members together with a majority of said members representing the United States vote in favor thereof; provided that any action not binding for such a reason may be ratified within thirty days by the concurrence in writing of a majority of the state members and the concurrence in writing of a majority of the members representing the United States.

Article VII: *Finances*

A. The commission shall submit to the governor or designated officer of each party state a request for funds to cover estimated expenditures for such period as may be required by the laws of that jurisdiction for presentation to the legislature thereof. Any such request shall indicate the sum or sums which the commission has requested or intends to request be appropriated by the United States for the use or support of the commission during the period covered thereby.

B. With due regard for such monies and other assistance as may be made available to it, the commission shall be provided with such funds by each of the several states participating therein to provide the means of establishing and maintaining facilities, a staff of personnel, and such activities as may be necessary to fulfill the powers and duties imposed upon and entrusted to the commission.

With due allowance for monies otherwise available, each budget of the commission shall be the responsibility of the party states, to be apportioned among them on a weighted formula based fifty percent on population and fifty percent on gross land area, such population and gross land area to be determined in accordance with the last official United States Census of Population, but provided that the total contributions of all of the states shall not be required to exceed fifty thousand dollars annually and provided further that regardless of the number of states party to the compact any time the maximum annual contribution required of any state shall not exceed its share of the fifty thousand dollars as determined above. Any state may contribute such funds in excess of its share, as determined above, as it may desire.

C. The commission shall not pledge the credit of any jurisdiction. The commission may meet any of its obligations in whole or in part with funds available to it under Article VIII(E) of this compact, provided that the commission takes specific action setting aside such funds prior to the incurring of any obligation to be met in whole or in part in such manner.

D. The members of this commission shall be paid by the commission their actual expenses incurred and incident to the performance of their duties, subject to the approval of the commission.

E. The commission shall keep accurate accounts of all receipts and disbursements. The receipts and disbursements of the commission shall be subject to the audit and accounting procedures established under its bylaws. However, all receipts and disbursements of funds handled by the commission shall be audited by a qualified public accountant and the report of the audit shall be included in and become a part of the annual report of the commission.

F. The accounts of the commission shall be open at any reasonable time for inspection by such agency, representative, or representatives of the jurisdictions which appropriate funds to the commission.

Article VIII: *Administration and Management*

A. The commission may sue and be sued, and shall have a seal.

B. The commission shall elect annually, from among its members, a chairman, vice-chairman and treasurer. The commission shall appoint an executive director who shall also act as secretary, and together with the treasurer, shall be bonded in such amounts as the commission may require.

C. The commission shall appoint and remove or discharge such personnel as may be necessary for the performance of its functions irrespective of any civil service laws which might otherwise apply. The commission shall establish and maintain, independently, by contract or agreement with the

United States or an agency thereof, or in conjunction with any one or more of the party states, suitable retirement programs for its employees. Employees of the commission shall be eligible for social security coverage in respect to old age and survivors insurance provided that the commission takes such steps as may be necessary pursuant to federal law to participate in such program of insurance as a governmental agency or unit. The commission may establish and maintain or participate in such additional programs of employee benefits as may be appropriate to afford employees of the commission terms and conditions of employment similar to those enjoyed by employees of the party states generally.

D. The commission may borrow, accept or contract for the services of personnel from any state or the United States or any subdivision or agency thereof, from any intergovernmental agency, or from any institution, person, firm or corporation.

E. The commission may accept for any of its purposes and functions under this compact any and all appropriations, donations, and grants of money, equipment, supplies, materials and services, conditional or otherwise, from any state or the United States or any subdivision or agency thereof, or intergovernmental agency, or any institution, person, firm or corporation, and may receive, utilize and dispose of the same.

F. The commission may establish and maintain such facilities as may be necessary for the transacting of its business. The commission may accept, hold, and convey real and personal property and any interest therein.

G. The commission may adopt, amend, and rescind by-laws, rules and regulations for the conduct of its business.

H. The commission shall make and transmit annually, to the legislature and governor of each party state, and to the President and Congress of the United States, a report covering the activities of the commission for the preceding year, and embodying such recommendations as may have been adopted by the commission. The commission may issue such additional reports as it may deem desirable.

Article IX: *Other Compacts and Activities*

Nothing in this compact shall be construed to impair, or otherwise affect, the jurisdiction of any interstate agency in which any party state participates, nor to abridge, impair, or otherwise affect the provisions of any compact to which any one or more of the party states may be a party, nor to supersede, diminish, or otherwise affect any obligation assumed under any such compact. Nor shall anything in this compact be construed to discourage additional interstate compacts among some or all of the party states for the management of natural resources, or the coordination of activities with respect to a specific natural resource or any aspect of natural resource management, or for the establishment of intergovernmental planning agencies in sub-areas of the region. Nothing in this compact shall be construed to limit the jurisdiction or activities of any participating government, agency, or officer thereof, or any private person or agency.

Article X: *Enactment*

A. This compact shall become effective when entered into and enacted into law by any three of the states of Connecticut, Maine, Massachusetts, New Hampshire, Rhode Island and Vermont, and when the United States has provided by law for the designation of its representation on the commission. Thereafter it shall become effective with respect to any other aforementioned state upon its enacting this compact into law.

B. Upon consent of the Congress of the United States of America, any other state in the northeastern area may become a party to this compact, by entering into and enacting this compact into law.

Article XI: *Withdrawal*

This compact shall continue in force and remain binding upon each party state until renounced by it. Renunciation of this compact must be preceded by sending three years' notice in writing of intention to withdraw from the compact to the governor of each of the other states party hereto and to such officers or agencies of the United States as may be designated by federal law.

Article XII: *Construction and Severability*

The provisions of this compact shall be severable and if any phrase, clause, sentence or provision of this compact is declared to be unconstitutional or the applicability thereof, to any state, agency, person, or circumstance is held invalid, the constitutionality of the remainder of this compact and the applicability thereof to any other state, agency, person or circumstance shall not be affected thereby. It is the legislative intent that the provisions of this compact be reasonably and liberally construed.

NEW ENGLAND GOVERNORS'
RESOLUTION – SEPTEMBER 18, 1965

Establishment of a New England River Basin Commission

WHEREAS, the New England Governors, with the concurrence of the Federal Inter-Agency Committee on Water Resources, have insured the conditions coordination of Federal and State programs over the past eight years through the establishment of a Northeastern Resources Committee; and

WHEREAS, New England has been a leader in the field of interstate cooperation, through interstate compacts, particularly with respect to water and other natural resources; and

WHEREAS, the Governors of the six states herein represented believe that New England would be a logical geographic and hydrologic unit under the program of water resources planning authorized by Public Law 89–80; and

WHEREAS, current drought conditions make it imperative that sound water resources planning and programming be intensified and accelerated without delay within the New England Region;

NOW, THEREFORE, be it resolved that the New England Governors, meeting in Newport, Rhode Island on 18 September 1965, do hereby request the Water Resources Council, under the provisions of Title II, Section 201 of the Water Resources Planning Act of 1965 (P. L. 89–80), to establish a New England River Basin Commission encompassing the area of the six New England States of Maine, New Hampshire, Vermont, Massachusetts, Connecticut, and Rhode Island, together with that portion of the State of New York within the drainage area of the Housatonic River, but specifically excluding those portions of the State of Vermont and Commonwealth of Massachusetts within the drainage area of the Hudson River, and that portion of the State of Vermont within the drainage area of Lake Champlain; and

BE IT FURTHER RESOLVED that each participating state contribute a sum not less than $5 thousand to cover the non-federal share of a New England River Basin Commission for the remainder of the current fiscal year.

BE IT FURTHER RESOLVED that the New England Governors shall immediately appoint their respective representatives to said Commission and that the Commission will become immediately operational and that said action shall not be dependent upon Federal funding.

SIGNED: John H. Reed
    Governor of Maine,
    Chairman, New England
    Governors' Conference

John W. King
Governor of New Hampshire

John A. Volpe
Governor of Massachusetts

John H. Chafee
Governor of Rhode Island

John Dempsey
Governor of Connecticut

Philip H. Hoff
Governor of Vermont

September 18, 1965.

PRESIDENTIAL EXECUTIVE ORDER – SEPTEMBER 6, 1967

Establishment of a New England
River Basins Commission (Executive Order 11371)

WHEREAS the Water Resources Planning Act (hereinafter referred to as the Act, 79 Stat, 244, 42 U.S.C. 1962 et seq.) authorizes the President to declare the establishment of a river basin water and related land resources commission when a request for such a commission is addressed in writing to the Water Resources Council (hereinafter referred to as the Council) by the Governor of a State within which all or part of the basin or basins concerned are located and when such a request is concurred in by the Council and by not less than one-half of the States within which portions of the basin or basins concerned are located; and

WHEREAS the Council, by resolution adopted October 14, 1965, concurred in the request of the Governor of the State of Maine as Chairman of the New England Governors' Conference, and did itself request that the President declare the establishment of the New England River Basin Commission under the provisions of section 201 of the Act; and

WHEREAS the request of the Governor of the State of Maine and the resolution of the Council of October 14, 1965, together with written concurrences by the Governors of the States of Maine, New Hampshire, Vermont, Massachusetts, Connecticut, Rhode Island, and New York, satisfy the formal requirements of section 201 of the Act; and

WHEREAS it appears that it would be in the public interest and in keeping with the intent of Congress to declare the establishment of such a Commission:

NOW, THEREFORE, by virtue of the authority vested in me by section 201 of the Act, and as President of the United States, it is ordered as follows:

SECTION 1. New England River Basins Commission. It is hereby declared that the New England River Basins Commission is established under the provisions of Title II of the Act.

SECTION 2. Jurisdiction of Commission. It is hereby determined that the jurisdiction of the New England River Basins Commission referred to in section 1 of this order (hereinafter referred to as the Commission) shall extend to the area of the six New England States of Maine, New Hampshire, Vermont, Massachusetts, Connecticut and Rhode Island, together with that portion of the State of New York within the drainage area of the Housatonic River, but specifically excluding those portions of the State of Vermont and Massachusetts within the drainage area of the Hudson River and excluding also that portion of the State of Vermont within the drainage area of Lake Champlain, in accordance with the request of the Governor of the State of Maine, concurred in by the Governors of the other New England States and New York, and in accordance with the resolution of the Council.

SECTION 3. Membership of the Commission. It is hereby determined that, in accordance with section 202 of the Act, the Commission shall consist of the following:

(1) a Chairman to be appointed by the President,
(2) one member from each of the following Federal departments and agencies: Department of Agriculture, Department of the Army, Department of Commerce, Department of Health, Education and Welfare, Department of Housing and Urban Development, Department of the Interior, Department of Transportation, and Federal Power Commission, each such member to be appointed by the head of each department or independent agency he represents,
(3) one member from each of the following States: Maine, New Hampshire, Vermont, Massachusetts, Connecticut, Rhode Island and New York, and
(4) one member from each interstate agency created by an interstate compact to which the consent of Congress has been given and whose jurisdiction extends to the waters of the area specified in section 2.

SECTION 4. Functions to be performed. The Commission and its Chairman, members, and employees are hereby authorized to perform and exercise, with respect to the jurisdiction specified in section of this order, the functions, powers, and duties of such a Commission and of such Chairman, members, and employees, respectively, as set out in Title II of the Act.

SECTION 5. International coordination. The Chairman of the Commission is hereby authorized and directed to refer to the Council any matters under consideration by the Commission which relate to the areas of interest or jurisdiction of the International Joint Commission, United States and Canada. The Council shall consult on these matters as appropriate with the Department of State and the International Joint Commission through its United States Section for the purpose of enhancing international coordination.

SECTION 6. Reporting to the President. The Chairman of the Commission shall report to the President through the Council.

SIGNED   LYNDON B. JOHNSON

The White House
September 6, 1967.

(Filed with the Office of the Federal Register, 3:20 p.m., September 7, 1967)

NEW ENGLAND GOVERNORS' RESOLUTION – NOVEMBER 16, 1967

Deactivation of the Northeastern Resources Committee

WHEREAS, there has been in New England, since the days of the New England–New York Inter-Agency Committee effort (1956–67), a most cooperative climate between the Federal and State interests in the broad and important field of Natural Resources; and

WHEREAS, as a result, on June 29, 1956, the Federal Inter-Agency Committee on Water Resources approved a charter for the formation and operation of the Northeastern Resources Committee; and

WHEREAS, this charter was approved by the New England Governors' Conference (composed of Governors Joseph B. Johnson, Vermont, Chairman; Edmund S. Muskie, Maine; Lane Dwinell, New Hampshire; Foster Furcolo, Massachusetts; Abraham A. Ribicoff, Connecticut; and Dennis J. Roberts, Rhode Island) on May 17, 1958; and

WHEREAS, on January 13, 1966, the Northeastern Resources Committee adopted a resolution that it be de-activated upon the establishment of the New England River Basins Commission; and

WHEREAS, on September 6, 1967, President Lyndon B. Johnson did, by Executive Order, establish the New England River Basins Commission and appointed R. Frank Gregg as chairman; and

WHEREAS, the New England River Basins Commission held its initial meeting in Boston, Massachusetts, October 16–17, 1967, and has perfected its organization and is ready to assume all obligations, duties and responsibilities imposed upon it under the Water Resources Planning Act of 1965 and by Executive Order; therefore

RESOLVED, That we the members of the New England Governors' Conference do hereby rescind our resolution of May 17, 1958, approving the establishment of the Northeastern Resources Committee and with our sincere thanks to the Northeastern Resources Committee, and do hereby agree their functions be taken over by the newly established New England River Basins Commission.

SIGNED: Philip H. Hoff
        Governor of Vermont,
        Chairman, New England
          Governors' Conference

John W. King
Governor of New Hampshire

Kenneth M. Curtis
Governor of Maine

John A. Volpe
Governor of Massachusetts

John H. Chafee
Governor of Rhode Island

John Dempsey
Governor of Connecticut

November 16, 1967.

A RESOLUTION OF THE STATE MEMBERS OF THE
NEW ENGLAND REGIONAL COMMISSION AND THE
NEW ENGLAND GOVERNORS' CONFERENCE
AFFIRMING THEIR COMMITMENT TO MAINTAIN A
REGIONAL FORUM FOR COORDINATION OF WATER
RESOURCES PLANNING AND MANAGEMENT IN
NEW ENGLAND

WHEREAS, the New England River Basins Commission, which was created by the President in 1967 at the request of the Governors of the New England states and New York, has provided a forum for coordinating the planning and management of water and related land resources by federal and state agencies and has helped to solve water problems common to all states; and

WHEREAS, the United States Water Resources Council is expected to vote to terminate the New England River Basins Commission, effective September 30, 1981; and

WHEREAS, the wise use, conservation and development of the region's water and related land resources continue to be important to the economic and social well being of its people; and

WHEREAS, the New England Governors since November 4, 1937, have been associated voluntarily for the purpose of assisting the New England states, Governors, Congress, and the Federal Executive Branch, and the public by conducting nonpartisan analysis study and research on their behalf in all matters of concern to these New England states, Governors, Congress, the Federal Executive Branch, and the public;

NOW THEREFORE BE IT RESOLVED that should the New England River Basins Commission be terminated, the State Members of the New England Regional Commission and the New England Governors' Conference agree to and affirm their commitment to maintain a regional forum for coordination of water resources planning and management in New England within the New England Governors' Conference for coordination of state and federal water resources planning and management; and

BE IT FURTHER RESOLVED that the State Members of the New England River Basins Commission are requested to provide for the transfer of all unobligated assets of the Commission at the time of its termination to a forum for coordination of water resources planning and management within the New England Governors' Conference; and

BE IT FURTHER RESOLVED that the State Members of the New England River Basins Commission are also requested to prepare recommendations, in consultation with the staff of the New England Governors' Conference, the Federal Members of the Commission and its citizen advisory committees, for consideration at the August, 1981, meeting of the Governors' Conference on the following matters:

• appropriate water resource program activities to be sponsored by the New England Governors' Conference;

- the role of relevant state (including New York) and federal agencies and the public in such activities;
- methods of securing continuing financial support for the water resource programs; and
- management, staffing and technical support for such activities.

This Resolution is effective immediately.

ADOPTION CERTIFIED BY THE STATE MEMBERS OF THE NEW ENGLAND REGIONAL COMMISSION AND THE NEW ENGLAND GOVERNORS' CONFERENCE ON JUNE 25, 1981

Hugh J. Gallen
Governor of New Hampshire
State Cochairman, New England
  Regional Commission
Chairman, New England Governors'
  Conference

PRESIDENTIAL EXECUTIVE ORDER – SEPTEMBER 9, 1981
TERMINATION OF RIVER BASIN COMMISSIONS

By the authority vested in me as President by the Constitution and laws of the United States, in order to ensure the orderly termination of the six river basin commissions established pursuant to the Water Resources Planning Act (42 U.S.C. 1962 *et seq.*), it is hereby ordered as follows:

Section 1. In accord with the decision of the Water Resources Council pursuant to Section 203(a) of the Water Resources Planning Act (42 U.S.C. 1962b-2(a)), the following river basin commissions shall terminate on the date indicated:

(a) Pacific Northwest River Basins Commission, terminated on September 30, 1981.
(b) Great Lakes Basin Commission, terminated on September 30, 1981.
(c) Ohio River Basin Commission, terminated on September 30, 1981.
(d) New England River Basins Commission, terminated on September 30, 1981.
(e) Missouri River Basin Commission, terminated on September 30, 1981.
(f) Upper Mississippi River Basin Commission, terminated on December 31, 1981.

Sec. 2. All Federal agencies shall cooperate with the commissions and the member States to achieve an orderly close out of commission activities and, if the member States so elect, to carry out an orderly transition of appropriate commission activities to the member States.

Sec. 3. To the extent permitted by law, the assets of the commissions which the Federal Government might otherwise be entitled to claim are to be transferred to the member States of the commissions, or such entities as the States acting through their representatives on the commissions may designate, to be used for such water and related land resources planning purposes as the States may decide among themselves. The terms and conditions for transfer of assets under this Section shall be subject to the approval of the Director of the Office of Management and Budget, or such Federal agency as he designates, before the transfer is effective.

Sec. 4. Federal agency members of river basin commissions are directed to continue coordination and cooperation in future State and inter-State basin planning arrangements.

Sec. 5. (a) Effective October 1, 1981, the following Executive Orders are revoked:

(1) Executive Order No. 11331, as amended, which established the Pacific Northwest River Basins Commission.
(2) Executive Order No. 11345, as amended, which established the Great Lakes Basin Commission.

(3) Executive Order No. 11371, as amended, which established the New England River Basins Commission.

(4) Executive Order No. 11578, as amended, which established the Ohio River Basin Commission.

(5) Executive Order No. 11658, as amended, which established the Missouri River Basin Commission.

(b) Effective January 1, 1982, Executive Order No. 11659, as amended, which established the Upper Mississippi River Basin Commission, is revoked.

SIGNED   RONALD REAGAN

THE WHITE HOUSE,
SEPTEMBER 9, 1981.

NEW ENGLAND GOVERNORS'
RESOLUTION – SEPTEMBER 18, 1981

A Resolution of the New England Governors' Conference
Further Defining the Regional Forum for the Coordination of
Water Resources Planning and Management in New England

WHEREAS, the Governors in their resolution number 247 requested the
State Members of the New England River Basins Commission to prepare
recommendations for consideration by the New England Governors' Conference on the structure of a regional forum for the coordination of water
resources planning and management in New England; and

WHEREAS, the New England River Basins Commission has been officially
terminated as of September 30, 1981; and

WHEREAS, participation of New York State in a regional water resources
forum has been useful and is desired.

NOW THEREFORE BE IT RESOLVED that the Governors approve the attached recommendations from the State Members of the New England
River Basins Commission for establishing a New England/New York Water
Council within the New England Governors' Conference except that the
Council is solely advisory to the Governors and the staff of the Council
shall be appointed by and shall report directly to the New England Governors' Conference; and

BE IT FURTHER RESOLVED that a continuing commitment to the support
of the New England/New York Water Council by the Governors' Conference is subject to the future availability of funding; and

BE IT FURTHER RESOLVED that New York State be invited to participate
formally in the New England/New York Water Council; and

BE IT FURTHER RESOLVED that the State Members of the New England
River Basins Commission are requested to prepare a plan further detailing a
program and budget for the operation of the New England/New York
Water Council in Fiscal Year 1982.

ADOPTION CERTIFIED BY THE NEW ENGLAND GOVERNORS' CONFERENCE on
September 18, 1981.

Hugh J. Gallen
Governor of New Hampshire
Chairman

Recommended Actions for Establishing a
New England/New York Water Council within the
New England Governors' Conference

The State Representatives of the New England River Basins Commission
agree on the following structure for maintaining a regional forum for the
coordination of water resources planning and management in New
England.

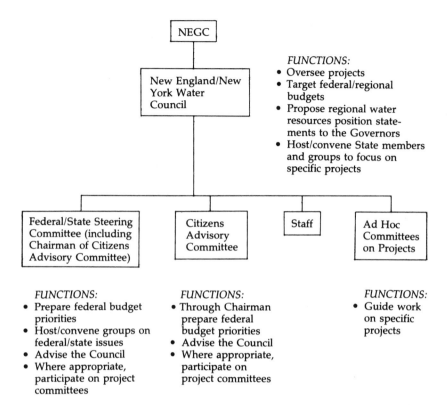

FUNCTIONS:
- Oversee projects
- Target federal/regional budgets
- Propose regional water resources position statements to the Governors
- Host/convene State members and groups to focus on specific projects

FUNCTIONS:
- Prepare federal budget priorities
- Host/convene groups on federal/state issues
- Advise the Council
- Where appropriate, participate on project committees

FUNCTIONS:
- Through Chairman prepare federal budget priorities
- Advise the Council
- Where appropriate, participate on project committees

FUNCTIONS:
- Guide work on specific projects

Underlying tenets:

1. The New England/New York Water Council, by virtue of its location within the New England Governors' Conference and its state funding, is a creature of the Governors. Accountability, therefore, will be to the elected political leaders of the New England states.
2. Its staff will be few in number and directly responsible to the Council. The staff will carry out the directives of and be sensitive to the interests of states.
3. We suggest that the State representatives comprising the members of the New England/New York Water Council be the people who were the State representatives to the New England River Basins Commission. As with NERBC, State members may be represented by alternates.

   - This body would permit continuity of knowledge, technical experience and contacts, as well as continue the links to the Governors.
   - Chairmanship of the Council will rotate and coincide with the Chairman of the Governors' Conference.

4. Federal agencies are invited to fully participate as equal partners in all areas of the Council's work, with the exception of formal ratification of policy positions recommended to the Governors.

5. The New England/New York Water Council be responsible for four functions:

a) *target federal funding for regional projects and programs.* Federal agencies feel that priority setting for regional funds is helpful to them and has had an influence in the distribution of available monies. They want this activity to continue after NERBC's demise.

b) *propose regional water resources position statements.* Governors need a technical arm to assist them in developing positions on critical water resources issues. Statements may respond to a federal initiative regarding legislation, policy or other federal activity or be generated within the region to influence federal, state, regional or local actions.

c) *host interactive discussions.* States need to maintain an official forum for exchanging ideas and dealing with emerging water resource issues. In addition, specific tasks may be better coordinated and facilitated if a regional forum convenes the affected parties.

To ensure that the discussions are both linked to the State's interests and to decision-makers who have the authority to implement discussion results, the topics will be approved by the Council's members.

d) *carry out projects.* Project work will be guided and managed by an advisory group established by and responsible to the Council members. Project staff will be accountable to the advisory group and hence to the Council and not to the permanent staff. Hopefully, this set-up will prevent the staff from developing their own constituency, separate from the States' interests.

On-going projects, which the States wish to continue, can be accommodated as they are currently established. The Coastal Zone Task Force, for example, would become the advisory group to the Coastal Zone Liaison Service.

Its staff person would continue to respond to the Task Force's interests.

6. Depending on the function, the involvement of federal officials and citizens will differ.

a) *Funding Priorities*

Federal agencies will be asked to participate as equal partners with states. Their commitment and attendance are essential if states are to influence the federal budgetary process.

The chairman of the Citizens Advisory Committee will also be invited to be a full participant. That person will be responsible for representing the interests of the many citizen committees and for reporting back to them. Because of limited Commission funds, no staff assistance can be offered. Although this will place a burden on citizens and particularly the chairman, the citizens who advised us in drafting our statement recognized this limitation and asked to be directly involved in the decision-making process. Their primary concern is to be heard.

b) *Position Statements*

The Citizen and Federal Advisory groups will be encouraged to advise the state members. The mechanisms for receiving this advice will neither be elaborate nor time-consuming and will depend on the situation. In some instances, the chairman may be asked to canvass opinions; in other cases, a larger group may be contacted.

c) *Hosting Sessions*

In the state exchanges the primary and, sometimes sole, participants will be the state representatives. At their request other interest groups can be included.

In the regional forums, affected and interested parties including citizens and federal officials will be full participants. The lead convener will be a state representative.

d) *Projects*

The sponsoring agency will be a full participant in the advisory group. The intention is to involve a full range of affected parties including federal, state, and local officials and citizen advocates. The make-up of each advisory group will reflect the specific project's characteristics.

NEW ENGLAND GOVERNORS'
RESOLUTION – SEPTEMBER 18, 1981

A Resolution of the New England Governors' Conference
Approving a Program and Budget for the New England/New York
Water Council, Making Initial Appointments to the Council,
and Accepting Funds from the New England River Basins Commission

WHEREAS, the New England Governors in previous resolutions have agreed to establish a forum for the coordination of water resources planning and management in New England because of the termination of the New England River Basins Commission and have approved the formation of a New England/New York Water Council to serve as such forum; and

WHEREAS, the New England River Basins Commission has voted to transfer its unobligated assets to the Governors' Conference for the New England/New York Water Council, subject to authorization by the President and approval by the Office of Management and Budget.

NOW THEREFORE BE IT RESOLVED that the Governors approve the attached recommendations from the State Members of the New England River Basins Commission for a program and budget for FY 1982 to serve as a guideline for use of funds transferred from NERBC; and

BE IT FURTHER RESOLVED that the Chairman of the New England Governors' Conference is authorized to enter into an agreement with the New England River Basins Commission providing for the transfer and use of the unobligated assets of NERBC in accordance with this Resolution; and

BE IT FURTHER RESOLVED that the current State Members and Alternates of the New England River Basins Commission are hereby named as the initial members and alternates of the New England/New York Water Council until such time as each Governor makes new appointments; and

BE IT FURTHER RESOLVED that the initial members of the New England/New York Water Council are requested to prepare recommendations for staffing and other matters necessary to provide for an orderly and timely initiation of the water resources program as soon as funds are available.

ADOPTION CERTIFIED BY THE NEW ENGLAND GOVERNORS' CONFERENCE on September 18, 1981.

Hugh J. Gallen
Governor of New Hampshire
Chairman

Recommended FY 82 Program and Budget for the
New England/New York Water Council of the New England
Governors' Conference

The State Members of the New England River Basins Commission propose the following program and budget for the New England/New York Water Council for FY 1982, to be used as a general guideline for use of funds and programs to be transferred from the New England River Basins Commission and to be further detailed after the exact amount of funds has been determined and transferred and the New England/New York Water Council has been established.

I. Program
  A. *Basic Program*
    1. For presentation to and consideration by the New England Governor's Conference:
      a. Prepare targets for federal regional budgets for FY 83/84
      b. Propose regional positions on current water resource issues to be selected by the Council. These might include hydropower (FERC exemptions, Power Act amendments, National Hydropower Study) and water supply (drought management, deteriorating infrastructure, ground water protection).
    2. Integrate and distribute reports on water supply and dredge management which are currently being prepared by each state under contract with NERBC.
    3. Oversee special projects, including management of the ongoing projects listed under B. below.
    4. Sponsor regional forums for the exchange of information and development of policy positions on selected water issues.
  B. *Ongoing Projects*
    Each of these projects is currently ongoing at NERBC with specifically earmarked funding.
    1. With separate funding from the coastal states, contingent upon receipt of federal CZM funds, support the coordination, information exchange and policy development activities of the New England/New York Coastal Zone Task Force. (See attached letter and resolution)
    2. Conduct OCS Information Program under the supervision of the Coastal Zone Task Force and in accordance with the attached FY 82 Work Program with surplus program funds transferred from NERBC, subject to approval of USGS.
    3. Continue Connecticut River Flood Education Program by providing flood plain management information to municipal officials and citizen groups with surplus program funds transferred from NERBC at the request of Massachusetts.
II. Budget—Estimated
  A. *For Basic Program*
    1. Income
       From NERBC—at least $165,000
    2. Expenditures
       Salaries and benefits   75,000
       Travel   5,000
       Operating Expenses   35,000
       Reserve for FY 83   50,000
                                          Total $165,000

    3. Staffing
       Two Professionals
       One Secretary

B. *For Coastal Zone Task Force*
    1. Income

From six coastal states	$ 30,000

    2. Expenditures

Salaries and benefits	27,000
Travel	3,000
Operating Expenses	(from basic program)
Total	$ 30,000

    3. Staffing
       One Professional (the Coastal Zone Liaison)

C. *For OCS Information Program*
    1. Income

From NERBC w/USGS approval	$ 45,000

    2. Expenditures

Salaries and benefits	23,000
Travel	2,000
Operating Expenses	20,000
Total	$ 45,000

    3. Staffing
       One Professional (OCS Information Specialist)

D. *For Connecticut River Flood Education Program*
    1. Revenue

from NERBC w/Massachusetts' approval	$ 9,000

    2. Expenditures

Operating Expenses	9,000

    3. Staffing
       None (support from basic program staff)

Notes

1. Introductory historical material from Time, Inc., *Time Capsule*, 1950.
2. Introductory legislative material from Shoemaker, *Conservation News*, National Wildlife Federation; and Shoemaker, *Conservation News Service*, Natural Resources Council of America.
3. Introductory material on the national setting from Maass, *Muddy Waters*, 1951; Leuchtenburg, *Flood Control Politics*, 1953; Smith and Castle, *Economics and Public Policy in Water Resource Development*, 1964; and Schad and Boswell, *Congressional Handling of Water Resources*, 1967.
4. Shoemaker, *Conservation News Service*, May 19, 1949.
5. Ibid., August 14, 1949.
6. Ibid., March 6, 1949.
7. Background material on the states from Leuchtenburg, *Flood Control Politics*, 1953; Boston Society of Civil Engineers, "Report of the Committee on Floods," 1930; interviews with state officials; state statutes and documents.
8. Background material on interstate compacts from Gere, *Patterns of Federal-Regional Interstate Cooperation in New England*, 1968; Leuchtenburg, *Flood Control Politics*, 1953; Shoemaker, *Conservation News*, 1949 and 1950; interviews with state and regional officials; state statutes and interstate documents.
9. Background material on economic problems from Council of Economic Advisers, *The New England Economy*, 1951; National Planning Association, *The Economic State of New England*, 1954; Donovan, "Watch the Yankees," *Fortune*, March, 1950; Handlin and Jones, "The Withering of New England," *Atlantic*, April, 1950; Kennedy, "New England and the South," *Atlantic*, January, 1954; New England Governors Conference, Minutes, 1945 and 1949; interviews and correspondence with regional officials; newspaper articles in *Washington Post, Hartford Times, Hartford Courant.*
10. Background material on hydroelectric power from New England Council, *Power in New England*, 1948; Federal Power Commission, *Power Market Survey*, 1949; Arris, *Water Power—Myth or Fact?*, 1949; Smith, *The Maine Power Problem*, 1951; Harris, *The Economics of New England*, 1952; Leuchtenburg, *Flood Control Politics*, 1953; New England Governors Conference, Minutes, 1951; New England Council, Minutes, 1945; interviews and correspondence with regional officials.
11. Background material on the region and the Congress from Schad and Boswell, *Congressional Handling of Water Resources*, 1967; *Congressional Quarterly Almanac*, 1950; Miller, *Henry Cabot Lodge*, 1967; interviews with regional officials.
12. Background material on the birth of NENYIAC from Shoemaker, *Conservation News, Conservation News Service*, 1949 and 1950; *Congressional Record*, 1949 and 1950; Leuchtenburg, *Flood Control Politics*, 1953; interviews and correspondence; newspaper articles in *Hartford Times, Hartford Courant, Portland Press-Herald.*

13. Leuchtenburg, *The New Leader*, March 12, 1949.
14. *Congressional Record*, March 14, 1949.
15. Letter from Philip Lagerquist, Harry S. Truman Library, Independence, Mo., February 14, 1968.
16. Letter from Philip Lagerquist, Harry S. Truman Library, Independence, Mo., February 14, 1968.
17. *Hartford Times*, October 14, 1950.
18. NENYIAC, Minutes, January 23, 1951.
19. Ibid.
20. Interview with William S. Wise, May 6, 1968.
21. NENYIAC, Minutes, January 23, 1951.
22. New England Council, New Hampshire State Dinner, November 15, 1951.
23. NENYIAC, Minutes, January 23, 1951.
24. Address by Col. Frederic F. Frech, Corps of Engineers, at the 25th Anniversary New England Conference, November 16, 1950.
25. *Hartford Courant*, January 27, 1951.
26. Interview with Anthony Wayne Smith, July 29, 1968.
27. NENYIAC, *Report of Testimony and Statements*, October 16, 1951.
28. Ibid., March 19, 1952.
29. Ibid., June 12, 1952.
30. Ibid., October 9, 1952.
31. *Congressional Record*, May 27, 1949.
32. NENYIAC, *Report of Testimony and Statements*, September 11, 1952.
33. Ibid., November 13, 1952.
34. NENYIAC, *Report of Testimony and Statements*, September 11, 1952.
35. Interview with Col. Gerald B. Troland (ret.), June 27, 1968.
36. Letter from Col. B. B. Talley to Lt. Gen. Lewis A. Pick, October 14, 1952.
37. Transition memorandum prepared by Leland Olds, December 19, 1952.
38. Letter from Mark Abelson to Lyle E. Craine, April 19, 1954.
39. Letter from Philip Lagerquist, Harry S. Truman Library, Independence, Mo., February 14, 1968.
40. NENYIAC Report, 1955. Part 1, Section I.
41. Memorandum from Col. B. B. Talley to Office of Chief of Engineers, April 23, 1954.
42. NENYIAC, *Report of Testimony and Statements*, November 9, 1954.
43. Ibid., November 10, 1954.
44. Ibid., December 14, 1954.
45. Ibid., December 15, 1954.
46. Ibid., December 16, 1954.
47. Ibid., January 12, 1955.
48. Ibid., January 13, 1955.
49. Ibid., February 8 and 10, 1955.
50. Ibid., March 10, 1955.
51. *Providence Journal-Bulletin*, March 7, 1955.
52. NENYIAC Executive Council, Minutes, March 10, 1955.
53. Letter from Col. B. B. Talley to Frederick Stueck, March 14, 1955.
54. NENYIAC Executive Council, Minutes, April 14, 1955.
55. Letter from George H. Roderick to Gov. Dennis J. Roberts, June 30, 1956.
56. New England Governors Conference, Minutes, February 25, 1957.
57. NRC, Minutes, April 11, 1957.

58. NRC, *Summation by Discussion Leaders* (New Hampshire Conference), 1958.
59. Interview with Ruth C. Matson, September 4, 1968.
60. NRC, Minutes, March 21, 1958.
61. NRC, Minutes, April 14, 1958.
62. NRC, Minutes, May 26, 1958.
63. House Public Works Committee, Staff Memorandum on H.R. 9999.
64. Letter from Rep. John W. McCormack to Gardner A. Caverly under date of June 13, 1960.
65. *Boston Globe*, August 3, 1961.
66. New England Governors Conference, Minutes, September 11, 1961.
67. Letter from Interior Secretary Stewart P. Udall to Sen. Dennis Chavez, January 11, 1962.
68. *Boston Globe*, September 19, 1962; *Concord Monitor*, September 18, 1962.
69. *Concord Monitor*, October 25, 1962.
70. NRC, Minutes, June 18, 1964.
71. New England Governors Conference, September 18, 1966.
72. *Providence Journal-Bulletin*, September 14, 1967.
73. NERBC, Minutes, October 16–17, 1967.
74. Ibid., May 20–21, 1969.
75. Meeting of river basin commission chairmen in Washington, D.C. November 20–21, 1969, as reported NERBC, Minutes, December 9–10, 1969.
76. Ibid.
77. NERBC, Minutes, Special meeting, February 9, 1971.
78. NERBC, Minutes, February 15–16, 1972.
79. Ibid., June 4–5, 1973.
80. Ibid., March 31, 1976.
81. Ibid., September 20–21, 1978.
82. Letter from John R. Ehrenfeld to NERBC members, February 17, 1981.
83. General Accounting Office, Report to the Congress, May 28, 1981.
84. Letter from Cdr. Stephen L. Richmond to Governor Edward J. King, May 15, 1981.
85. Ibid.
86. Letter from Cdr. Stephen L. Richmond to Secretary of the Interior James Watt, September 22, 1981.
87. Open letter to the citizens of New England/New York from Comdr. Stephen L. Richmond, September 28, 1981.
88. Interview with John McAleer, Washington, D.C., May 27, 1968.
89. Interview with Harold F. Schnurle, Augusta, Me., August 16, 1968.
90. Interview with Edwin W. Webber, Washington, D.C., January 31, 1968.
91. Interview with Thomas J. Rouner, Westmoreland, N.H., May 11, 1968.
92. Interview with Arthur Ristau, Burlington, Vt., November 11, 1981.
93. Interview with John R. Ehrenfeld, Boston, November 4, 1981.
94. Interview with Peter Piattoni, Concord, N.H., November 22, 1981.
95. Interview with Comdr. Stephen L. Richmond, Boston, November 4, 1981.
96. Wildlife Conservation, Inc., *Conservation Currents*, December 1954.

Personal Interviews

Mark Abelson
Northeast Regional Coordinator
U.S. Department of the Interior

A. William Albert
Director, Division of Water
 Pollution Control
Vermont Department of Water
 Resources

David Arnold
Executive Director
Council of Northeast Governors

George H. Arris
Financial Editor
Providence (RI) Journal-Bulletin

Richard Barringer
Director
Maine State Planning Office

Daniel Beard
Executive Director
Renewable Natural Resources
 Foundation

Bernard B. Berger
Graduate Center
University of Massachusetts

Paul O. Bofinger
President
Society for the Protection of New
 Hampshire Forests

Justin Brande
Adjunct Professor of
 Environmental Studies
University of Vermont

Richard Brett
Vermont Natural Resources Council

Robert D. Brown
Staff Director
New England River Basins
 Commission

Arthur W. Brownell
International Paper Company

Paul E. Bruns
Chairman, Department of Forest
 Resources
University of New Hampshire

William R. Burch, Jr.
Professor of Forest Sociology
Yale University

John P. Chandler
Col. (Retired)
U.S. Army Corps of Engineers

Anthony S. Codding
Director
New England Center for
 Continuing Education

Charles W. Colson
Washington Counsel
New England Council

John Crossman
President
Connecticut River Watershed
 Council

Kenneth M. Curtis
Former Governor of Maine

Dennis W. Ducsik
Associate Professor, Center for
 Technology, Environment and
 Development
Clark University

Calvin B. Dunwoody
Chief, Division of Planning &
 Development
Rhode Island Department of
 Environmental Management

Paul Eastman
Executive Director
Interstate Commission on
 Potomac River Basin

John R. Ehrenfeld
Senior Staff Member
Arthur D. Little, Inc.

Robert W. Eisenmenger
Senior Vice President and
 Director of Research
Federal Reserve Bank of Boston

George H. Ellis
President
Federal Reserve Bank of Boston

Stephen Ells
Office of Intergovernmental
 Liaison
U.S. Environmental Protection
 Agency

John A. Finck
Director of Water Resources
 Planning
New York Department of
 Environmental Conservation

Joseph L. Fisher
Director, Economic Policy
 Department
Wilderness Society

Rockwood H. Foster
Member, C. & O. Advisory
 Commission
U.S. Department of the Interior

Robert J. Frazier
Chief, Comprehensive Basin
 Planning Section
Connecticut River Comprehensive
 Study
U.S. Army Corps of Engineers

Joel Frisch
Deputy Director, Regional
 Programs Division
U.S. Water Resources Council

Jack Frost
Assistant Director, Office of
 Operations
U.S. Water Resources Council

Robert H. Gardiner
Executive Director
Maine Natural Resources Council

Margaret Garland
Director
Vermont State Energy Office

Alec Giffen
Planner
Maine State Planning Office

Thomas Glenn
Executive Director
Interstate Sanitation Commission

Ernest M. Gould, Jr.
Forest Economist
Harvard Forest

R. Frank Gregg
School of Renewable Natural
 Resources
University of Arizona

Gilbert Gude
Director
Congressional Research Service

Bartlett Hague
Chief of Environmental Studies,
 Water Quality Branch
U.S. Environmental Protection
 Agency

Mary Louise Hancock
Former State Senator
  (New Hampshire)

Rudolph Hardy
Executive Director
New England Economic Research
  Foundation

Robert N. Haskell
President
Bangor (Maine) Hydro-Electric
  Company

Alfred L. Hawkes
Executive Director
Audubon Society of Rhode Island

John Hibbard
Secretary – Forester
Connecticut Forest and Park
  Association

David L. Holmes
Director of Comprehensive
  Planning
City of New Haven (Connecticut)

John D. Hull
U.S. Soil Conservation Service

Joseph Ignazio
Chief of the Planning Division
  (New England)
U.S. Army Corps of Engineers

Martin Inwald
Chief Civil Engineer
Federal Energy Regulatory
  Commission

Lloyd Irland
State Economist
Maine State Planning Office

Bernard B. Johnson
Assistant Director
Vermont State Planning Office

Elizabeth Kline
Assistant Secretary
Massachusetts Executive Office
  of Environmental Affairs

Charles E. Knox
District Engineer, Surface Water
  Branch
U.S. Geological Survey

Joseph C. Knox
Associate Sanitary Engineer
Camp, Dresser and McKee

Glenn Kumekawa
Director, Department of
  Intergovernmental Analysis
University of Rhode Island

Henry Lee, Jr.
Director, Office of Energy and
  Environment
Kennedy School of Government
Harvard University

Arthur Maass
Professor, Kennedy School of
  Government
Harvard University

John McAleer
Office of the Chief of Engineers
U.S. Army Corps of Engineers

Richard Macomber
Board of Rivers and Harbors
U.S. Army Corps of Engineers

Ruth Matson
Executive Secretary
Merrimack River Valley Flood
  Control Commission

Rev. Jerry Moore, Jr.
Chairman, Committee on
  Transportation and
  Environmental Affairs
District of Columbia City Council

Lois Murray
Board Member
New England Rivers Center

Walter Newman
Consultant
U.S. Environmental Protection
 Agency

Robert W. Patterson
Maine Natural Resources Council

Melvin Peach
Manager
New England Council

Alfred E. Peloquin
Executive Secretary
New England Interstate Water
 Pollution Control Commission

Peter Piattoni
Planner, New Hampshire State
 Planning Office
Chairman, New England–New
 York Water Council

Raymond Powell
Public Information Officer
U.S. Water Resources Council

Carl H. Reidel
Director, Environmental Program
University of Vermont

Commander Stephen L. Richmond
Planning Officer, First District
U.S. Coast Guard

Stephen Riley
Executive Director
Massachusetts Historical Society

Arthur Ristau
Green Mountain Power
 Company

Thomas J. Rouner
Vice President
New England Power Company

Frederick Sargent
Professor of Regional Planning
 and Resource Economics
University of Vermont

Theodore M. Schad
Deputy Executive Director
Commission on Natural Resources
National Academy of Science/
 National Research Council

Melvin E. Scheidt
Lecturer in Water Resources
Johns Hopkins University

Frederick Schmidt
Acting Director, Rural Studies
 Center
University of Vermont

Harold F. Schnurle
Vice President
Central Maine Power Company

Harry Schwarz
Lecturer in Water Resources
Clark University

Gerald Seinwill
Director
U.S. Water Resources Council

Henry Silbermann
Assistant Secretary
Maryland Department of Natural
 Resources

Anthony Wayne Smith
President
National Parks and Conservation
 Association

H. Bailey Spencer
Executive Director
New England Congressional
 Caucus

Harry A. Steele
Assistant Director for Planning
U.S. Water Resources Council

Chapman Stockford
Executive Director
Northeast Public Power
  Association

Charles Stoddard
Regional Coordinator, Upper
  Mississippi–Great Lakes Area
U.S. Department of the Interior

Frank Thomas
Acting Director
U.S. Water Resources Council

Hugo Thomas
Director, Natural Resources Center
Connecticut Department of
  Environmental Protection

Gerard B. Troland
Colonel (Retired)
U.S. Army Corps of Engineers

Daniel W. Varin
Chief
Rhode Island Statewide Planning
  Program

Warren Viessman
Senior Specialist in Engineering
  and Public Works
Congressional Research Service

Alvin C. Watson
Executive Secretary
Potomac Basin Advisory
  Committee

Edwin W. Webber
Director, Office of Area Planning
  and Program Support
U.S. Economic Development
  Administration

Seward Weber
Executive Director
Vermont Natural Resources
  Council

Mitchell Wendell
Chief Counsel
Council of State Governments

Brendan Whittaker
Secretary
Vermont Agency of
  Environmental Conservation

Austin H. Wilkins
Commissioner
Maine Forestry Department

Leonard U. Wilson
Consultant
Council of State Governments

William S. Wise
Director
Connecticut Water Resources
  Commission

Albert C. Worrell
Professor of Forest Policy
Yale University

Selected Readings

Allee, David, Leonard B. Dworsky, and Ronald North, October 1981. "U.S. Water Planning and Management, I: A Preliminary Report Prepared for the American Water Resources Association." Cornell University, Ithaca, N.Y.

Arris, George H., 1949. *Water Power: Myth or Fact?* Providence Journal Company. Providence, R.I.

Berger, Bernard B., Madge O. Ertel, and Edward R. Kaynor, May 1980. "Integrated Management of the Connecticut River Basin." Paper presented to the Conference on Unified River Basin Management, American Water Resources Association. Water Resources Research Center, University of Massachusetts. Amherst, Mass.

Boston Society of Civil Engineers, September 1930. Report of the Committee on Floods, March 1930. *Journal of the Boston Society of Civil Engineers* (XVII). Boston.

Bowen, David Karr, 1979. "Federal Activities To Be Covered under a Revised Water Resources Council Consistency Policy Exclusion Criteria." Report submitted to the U.S. Water Resources Council. Washington, D.C.

Bradley, Phillips, June 1935. "A TVA for New England?" *The American City* (V). New York.

Chesapeake Research Consortium, October 1977. Proceedings of the Bi-State Conference on the Chesapeake Bay. Virginia Institute of Marine Science.

Clark, Jon, et al., July 1981. "Building an Energy Consensus: Key Issues for the Eighties." Northeast-Midwest Institute. Washington, D.C.

Collins, Carrie, April 7, 1981. Briefing Paper on Water Distribution Systems and the Problem of Deterioration. New England Congressional Caucus. Washington, D.C.

Consortium of Northeast Organizations, September 1979. "Water Resource Priorities for the Northeast: The Northeast Water Resources Project." Report prepared by the Nova Institute. Northeast-Midwest Institute. Washington, D.C.

Council of Economic Advisers, 1951. "The New England Economy." Report of the Committee on the New England Economy. Washington, D.C.

Council of State Governments, December 1977. "Interstate compacts 1783–1977." Revised compilation. Lexington, Ky.

Curtis, Kenneth M. and Kermit Lipez, October 1975. "An Evaluation of Proposals for Regional Institutional Arrangements for Natural Resources Planning in New England." New England River Basins Commission. Boston.

Dean, Alan L., 1957. "Advantages and Disadvantages in the Use of Ad Hoc Commissions for Policy Formulation." Paper prepared for the annual meeting of the American Political Science Association, New York, September 5–7, 1957.

_____, 1957. "The President's Water Resources Policy Commission." Paper prepared for the annual meeting of the American Political Science Association, New York, September 5–7, 1957.

217

218

Derthick, Martha, 1974. *Between state and nation*. Brookings Institution. Washington, D.C.

Dishman, Robert B., and George Goodwin, Jr., September 1967. "State Legislatures in New England Politics: Final Report of the New England Assembly on State Legislatures." New England Center for Continuing Education. Durham, N.H.

Dodge, James, 1981. "Living by Life: Some Bioregional Theory and Practice." *Coevolution Quarterly* (Winter 1981).

Ducsik, Dennis W., April 1981. "Citizen Participation in Power Plant Siting: Aladdin's Lamp or Pandora's Box." Reprint No. 19. Center for Technology, Environment and Development. Clark University. Worcester, Mass.

———, April 28, 1981. "Citizen Participation in Power Plant Siting: The Utility Experience." Paper prepared for the American Planning Association National Conference, Boston. Clark University. Worcester, Mass.

Dworsky, Leonard B., and David J. Allee, July 1980. "Potential Interstate Institutional Entities for Water Resource Planning: A Report to the U.S. Water Resources Council." Cornell University. Ithaca, N.Y.

Ehrenfeld, John R., October 23, 1979. "The Role of River Basin Commissions in Comprehensive Water Resource Planning." Remarks at the USA/USSR Symposium on River Basin Water Quality, Planning and Management. Cambridge, Mass.

Engineers Joint Council, July 1951. "Principles of a sound water policy." Report of the National Water Policy Panel of the Engineers Joint Council. New York.

Fields, Ralph M. Associates, November 1979. "Regional Water Resource Management Planning: A Review of Level B Study Impacts." U.S. Water Resources Council. Washington, D.C.

———, January 1980. "Regional Water Resource Management Planning: Improving the U.S. Water Resources Council Comprehensive Studies Program (Level B Studies)." Water Resources Council. Washington, D.C.

———, October 19, 1981. "Regional and River Basin Level B Studies: A Summary Report." U.S. Water Resources Council. Washington, D.C.

Frisch, Joel, and Thomas Chaney, October 1981. "Current Status of the Replacements for Title II River Basin Planning Commissions." U.S. Water Resources Council. Washington, D.C.

Fesler, James W. (ed.), September 1950. "Government and Water Resources." *American Political Science Review* (XLIV). Washington, D.C.

Fox, Irving K., Summer 1957. "National Water Resources Policy Issues." *Law and Contemporary Problems* XXII (3), School of Law, Duke University. Durham, N.C.

Gere, Edwin A., 1968. "Patterns of Federal-Regional Interstate Cooperation in New England." Graduate School of Public Affairs, State University of New York. Albany.

Hartley, David K., January 14, 1981. "A Revised WRC Consistency Policy: A Consultant Report." Water Resources Council. Washington, D.C.

Hoover Commission Report on Organization of the Executive Branch of the Government, 1949. McGraw-Hill. New York.

House Committee on Public Works, 86th Congress, 2nd Session. Hearings on H.R. 9999 and H.R. 10022 (Northeastern Water and Related Land Resources Compact) held in Washington, D.C. March 30–31, 1960.

———, 87th Congress, 1st Session. Report to accompany H.R. 30 (Northeastern Water and Related Land Resources Compact). House Report 707. July 12, 1961.

Kennedy, John F., January 1954. "New England and the South: The Struggle for Industry." *Atlantic Monthly* (CXCIII). Boston, Mass.

Koch, Stuart G., 1981. *Water Resources Planning in New England.* University Press of New England. Hanover, N.H.

Kracke, Ernesta H., June 1978. "State Growth Policies in New England: Their Status, Successes and Shortcomings." Report prepared for the New England Natural Resources Center and the Rockefeller Foundation. New England Natural Resources Center. Boston.

Krutilla, John V. and Otto Eckstein, 1958. *Multiple-Purpose River Development: Studies in Applied Economic Analysis.* Johns Hopkins University Press. Baltimore, Md.

Leach, Richard H., and Redding S. Sugg, Jr., 1959. *The Administration of Interstate Compacts.* Louisiana State University Press. Baton Rouge.

League of Women Voters Education Fund, November 1967. *An Introduction to Comprehensive River Basin Planning: Structure and Strategy.* Washington, D.C.

Leuchtenburg, William E., 1953. *Flood Control Politics: The Connecticut River Valley Problem, 1927–1950.* Harvard University Press. Cambridge, Mass.

Little, Arthur D., Inc., 1964. "Projective Economic Studies of New England." Report prepared for the U.S. Army Engineer Division, New England, Corps of Engineers. Cambridge, Mass.

Lockard, Duane, 1959. *New England State Politics.* Princeton University Press. Princeton, N.J.

Maass, Arthur, 1951. *Muddy Waters.* Harvard University Press. Cambridge, Mass.

Martin, Roscoe C., Guthrie S. Birkhead, Jesse Burkhead, and Frank J. Munger, 1960. *River Basin Administration and the Delaware.* Syracuse University Press. Syracuse, N.Y.

Morgan, Stephen J., August 14, 1981. "Evaluating the New England Energy Congress." New England Congressional Institute. Washington, D.C.

National Oceanic and Atmospheric Administration, September 15, 1976. "Fishery Conservation and Management: Interim Regulations." U.S. Department of Commerce. Washington, D.C.

National Planning Association, 1954. *The Economic State of New England. Report of the Committee of New England.* Yale University Press. New Haven, Ct.

New England Center for Continuing Education, Annual Report 1979–80. Durham, N.H.

New England Council, 1948. "Power in New England." Report of the Power Survey Committee. Boston.

New England Interstate Water Pollution Control Commission. Annual reports for fiscal years 1978 and 1979. Boston.

New England Energy Congress. "A Blueprint for Energy Action" Final report, executive summary and recommendations, May 1979. Tufts University. Somerville, Mass.

New England Regional Commission, 1981. *The New England Regional Plan: An Economic Development Strategy.* University Press of New England. Hanover, N.H.

New England River Basins Commission, Annual reports for fiscal years 1968–1981. Boston.

———, Minutes of meetings 1968–1981. Boston.

220

_____, September 14, 1972 (revised December 6, 1972). "Strategies for Natural Resources Decision-Making." Boston.

_____, October 1979. "Predicting Impacts in the Connecticut River System: Executive Summary." Boston.

_____, September 1981. "NERBC Committees and Task Forces: A Directory of People with Information about New England's Water Resources." Boston.

Northeastern Resources Committee, January 1967. "Potential Program for New England River Basins Commission." Boston.

Pacific Fishery Management Council, 1977–78. Progress Report. Portland, Oreg.

Peirce, Neal R., 1976. *The New England States: People, Politics and Power in the six New England states*. W. W. Norton. New York.

Records of the Federal Inter-Agency River Basins Committee. Water Resources Council. Washington, D.C.

Records of the Inter-Agency Committee on Water Resources. Water Resources Council. Washington, D.C.

Records of the New England Council. New England Council. Boston.

Records of the New England Governors Conference. New England Governors Conference. Boston.

Records of the New England–New York Inter-Agency Committee. U.S. Army Corps of Engineers, New England Division. Waltham, Mass.

Records of the New England River Basins Commission. U.S. Army Corps of Engineers, New England Division. Waltham, Mass.

Records of the Northeastern Resources Committee. U.S. Army Corps of Engineers, New England Division. Waltham, Mass.

Records of the Water Resources Council. Water Resources Council. Washington, D.C.

Report of the New England–New York Interagency Committee, 1957. "Land and Water Resources of the New England–New York region." Senate Document 14, 85th Congress, 1st Session.

Report of the President's Advisory Committee on Water Resources Policy, 1956. House Document 315, 84th Congress, 2nd Session.

Report of the President's Water Resources Policy Commission, 1950. *A Water Policy for the American people*. Volume I of the report. Washington, D.C.

Report of the President's Water Resources Policy Commission, 1950. *Ten Rivers in America's Future*. Volume II of the report. Washington, D.C.

Russell, Howard S., 1981. *A Long, Deep Furrow*. University Press of New England. Hanover, N.H.

Schad, Theodore M., and Elizabeth M. Boswell, December 15, 1967. "Congressional Handling of Water Resources." Legislative Reference Service. Library of Congress. Washington, D.C.

Schad, Theodore M., and Elizabeth M. Boswell, 1968. "History of the Implementation of the Recommendations of the Senate Select Committee on National Water Resources." Legislative Reference Service. Library of Congress. Washington, D.C.

Senate Committee on the Judiciary, 87th Congress, 2nd Session. Hearings on H.R. 30 (Northeastern Water and Related Land Resources Compact) held in Washington, D.C. September 18, 1962.

Senate Committee on Public Works, 81st Congress, 2nd Session. Hearings on S. 3707 (New England–New York Resources Survey Commission) held in Washington, D.C. June 29, 1950 and July 5, 1950.

Shanley, Robert A. (ed.), July 1965. "Intergovernmental Challenges in New England." Bureau of Government Research. University of Massachusetts. Amherst, Mass.

Shea, Cordelia, March 7, 1979. "New England Water Briefing Paper." New England Congressional Caucus. Washington, D.C.

Smith, Stephen C., and Emery N. Castle (eds.), 1964. *Economics and Public Policy in Water Resource Development.* Iowa State University Press. Ames.

Southern Governors Association, 1980–81. Minutes of meetings of the Association and its Staff Advisory Committee. Council of State Governments. Atlanta, Ga.

_____, February 22, 1981. Report on regional coordination and cooperation. Council of State Governments. Atlanta, Ga.

Subcommittee on Water Resources, May 5–6, 1981. Extension of Authorization for the Water Resources Council (Hearings). House Committee on Public Works and Transportation. Washington, D.C.

Sullivan, Michael, and Richard Durkin, September 1980. "Water Resource Priorities for the Midwest: The Midwest Water Resources Project." Northeast-Midwest Institute. Washington, D.C.

Time Capsule, 1950. A history of the year condensed from the pages of *Time.* Time Inc. New York.

UNESCO, 1958. "Integrated River Basin Development. Report by a Panel of Experts." New York.

U.S. Advisory Commission on Intergovernmental Relations, April 1972. "Multistate Regionalism: A Commission Report." Washington, D.C.

U.S. Comptroller General, August 22, 1980. "Federal-State Environmental Programs—The State Perspective: A Compilation of Questionnaire Responses." General Accounting Office. Washington, D.C.

_____, November 13, 1980. "Congressional Guidance Needed on Federal Cost Share of Water Resource Projects When Project Benefits Are Not Widespread." General Accounting Office. Washington, D.C.

_____, February 20, 1981. "Federal-Interstate Compact Commissions: Useful Mechanisms for Planning and Managing River Basin Operations." General Accounting Office. Washington, D.C.

_____, May 28, 1981. "River Basin Commissions Have Been Helpful, but Changes Are Needed." General Accounting Office. Washington, D.C.

U.S. Department of the Interior, June 6, 1980. Final Report on Phase 1 of Water Policy Implementation. Report submitted to the President by the Secretary of the Interior. Washington, D.C.

U.S. General Accounting Office, October 31, 1977. "Improvements Needed by the Water Resources Council and River Basin Commissions to Achieve the Objectives of the Water Resources Planning Act of 1965." Washington, D.C.

_____, December 30, 1980. "Environmental Protection Issues in the 1980s." Washington, D.C.

U.S. Water Resources Council, July 2, 1980. "Improving the Planning and Management of the Nation's Water Resources." Washington, D.C.

_____, July 30, 1980. "104 Review of Regional Water Resource Management Plans." Washington, D.C.

_____, June 18, 1981. "Historical Highlights of the U.S. Water Resources Council's Principles, Standards, and Procedures and Level B Studies in Water and Related Land Resources Planning. In draft. Washington, D.C.

———, September 1981. "Regional Water Resource's Management Plan Review." Washington, D.C.

Viessman, Warren, Jr. and Christopher Caudill, September 1975. "The Water Resources Planning Act: An Assessment." Report of the Subcommittee on Energy Research and Water Resources of the Senate Committee on Interior and Insular Affairs. Washington, D.C.

Viessman, Warren, Jr., June 1978. "The Water Resources Policy Study: An Assessment." Report prepared for the Senate Committee on Energy and Natural Resources. Congressional Research Service. Washington, D.C.

———, November 22, 1978. "Coordination of Federal Water Resources Policies and Programs." Congressional Research Service. Washington, D.C.

———, September 1978. "An Analysis of the President's Water Policy Initiatives." Report prepared for the Senate Committee on Energy and Natural Resources. Congressional Research Service. Washington, D.C.

———, June 1979. "Priority Setting for Water Resources Projects and Programs." Report prepared for the Subcommittee on Water Resources of the Senate Committee on Environment and Public Works. Congressional Research Service. Washington, D.C.

———, December 1980. "Assessing the Nation's Water Resources: Issues and Options." Report prepared for the Senate Committee on Environment and Public Works. Congressional Research Service. Washington, D.C.

Wengert, Norman, Spring 1957. "The Politics of River Basin Development." *Law and Contemporary Problems* XXII (2). School of Law, Duke University. Durham, N.C.

Wengert, Norman and John C. Honey, 1954. "Program Planning in the U.S. Department of the Interior, 1946–53." *Public Administration Review* XIV (3). Chicago.

Western States Water Council, November 1974. "Western States' Water Requirements for Energy Development to 1990." Salt Lake City, Utah.

———, May 1, 1977. "Western States' Water Resource Organizations: Future Roles." Report prepared for the Western Governors. Salt Lake City, Utah.

———, Annual Report 1980. Salt Lake City, Utah.

———, November 1981. "State/Federal Financing and Western Water Resource Development." Salt Lake City, Utah.

Wilson, Leonard U., 1977. "State Strategies for Multistate Organizations." State Planning Series 8. Council of State Planning Agencies. Washington, D.C.

Wilson, Leonard U. and Anne D. Stubbs, August 1981. "The Commission's Reach: A Requiem Appraisal." Council of State Governments. Lexington, Ky.

Wilson, Leonard U., September 1981. "Water Resource Management: New Responsibilities for State Governments." Report of the Working Group on Changing Directions of State Water Agencies. Council of State Governments. Lexington, Ky.

Wright, John K. (ed.), 1933. "New England's Prospect: 1933." Special Publication No. 16, American Geographical Society. New York.

Zimmerman, Frederick L., and Mitchell Wendell, 1951. "The Interstate Compact since 1925." Council of State Governments. Chicago.

Index